What Are Consciousness, Animacy, Mental Activity and the Like?

Vladimir A. Lefebvre

Leaf & Oaks Publishers

Los Angeles

First publication Cogito-Centre, 2013 (in Russian)
Translation by Victorina D. Lefebvre

The book describes how the representation of certain psychological processes with a chain of heat engines allows modeling a large number of mental phenomena. This relation seems enigmatic in so far as we cannot find any analogues to heat engines in human body. Nevertheless, in his previous book, *The Cosmic Subject*, the author tried to link the chains of heat engines with objects physically existing in the reality. In his new book, the author chose a fundamentally different way. He revises the very concept of *existing* by considering that the ideal physical objects may also have a status of real existence. Mental phenomenology, according to the author, is a manifestation of existence of such ideal objects. Along this way the author deduces the main psychophysical laws and the set of harmonical intervals in music.

Key words: subjectivity, consciousness, awareness, Plato, bipolar choice, golden section, categorization, thermodynamics, musical intervals

ISBN   978-0-578-14136-7

CONTENT

To Vicka

Preface

How do we know which objects are animate and which inanimate? Are computers and robots animate? Some researchers would say that they are nothing more than electronic circuits, but others would assure us that elementary animacy certainly exists in robots. These opposite opinions reflect a fundamental scientific problem: What does it mean to say that a physical body is animate? What test can be conducted to show whether an object is animate?

The term *animacy* implies religious context. Animate objects are those into which the Creator has breathed a soul. Scientists can accept such a notion only if they are able to photograph the Creator at the moment of endowing matter with soul. Science is based on observable evidence rather than on belief. In this book I will adhere to the scientific approach.

I should note that the terms *animacy* and *consciousness* are often used as synonyms. Of course, in doing so we assume that even a spider has some rudiments of consciousness. Also, the words *animate* and *living* are close in meaning; in some languages, a single word comprises both meanings. I will treat these words as synonyms.

My approach is as follows. First, I will devise a philosophy such that consciousness may exist, but without assuming the agency of a Creator. Plato's ontology is essential to this philosophy.

Second, I will propose a hypothesis and test it empirically and experimentally. This testing has taken many years; I have used the work of many different researchers. The analysis of their work is scattered throughout different papers and books of mine. I consider it important to bring these published analyses together in compact form.

The book has three Appendices. In the first one, an introduction to thermodynamics is described; this material is similar to what was published earlier in my book, *The Cosmic Subject*. The second appendix contains a paper from 1965, in which my study of this subject began, and the third one is my paper on animal behavior co-written with Federico Sanabria.

Introduction

Within the last thirty years the study of consciousness has changed radically. Most researchers in the fields of psychology, physiology, and general systems studies used to believe that the mystery of consciousness would be revealed through better understanding of brain structure. Today this belief is no longer generally held (Myin, 2010). John Eccles (1979) was among the first physiologists to publicly take this position. After many years of research, he concluded that the human brain is connected with some hidden entity that determines whether neurons are on or off. Eccles suggested that this unobservable essence is consciousness.

Attitudes have changed within philosophy as well. David Chalmers (1996) introduced the metaphor of light to the problem of consciousness. Many suppose that consciousness is the result of a very complicated physical process in the brain. Why then, if we liken consciousness to light, should this process not occur in darkness? Where does the light come from? Why do we feel? Chalmers distinguished between easy problems and the "difficult problem." Finding functional and structural links between elements in our mental experience is an easy problem. The difficult problem is this: How do we understand and explain mental experience itself?

In this book I attempt to outline an approach toward solving the *difficult* problem. To do so, I will use Plato's ontology, which differs notably from today's natural-science ontology. In Plato's scheme, the phenomenon of consciousness is an essential component of everything that exists. A quantity of empirical evidence can be presented in favor of this ontology.

My first paper on this topic was published in 1965 (reprinted in this book as Appendix II). I suggested there that the process of self-organization is connected with an element-designer providing the system with structure and playing the role of the reflexion within the system, i.e., acting as its consciousness (Lefebvre, 1965).

— · —

I am very grateful to Victorina Lefebvre for her criticism and numerous valuable suggestions. Without her help and support, this work would have not been completed. I am also thankful to Harold Baker for linguistic advice and corrections.

Chapter 1.
The Ontology of Consciousness

In beginning to examine animacy, consciousness, mental experience, and the like, we feel a certain impotence in having to rely on metaphors rather than on clear objective signs of the phenomena. In this Chapter we will revise and broaden our concept of reality such that mental phenomena are included together with physical objects.

1.1. Clarifying the problem

Imagine that an electronics company has created a super-robot to work on Mars. Its appearance does not differ from that of a human being. The robot is supposed to spend two years on Mars, after which its batteries become exhausted and it dies. Imagine next that the robot says publicly that it does not want to fly to Mars, and that the thought of dying causes unbearable suffering. As a result, a mass movement arises in its defense, advocating that the robot be saved from this anguish. The company that created the robot publishes a document claiming that, unlike human beings and animals, robots cannot experience mental pain, and that the robot's "rebellion" is the result of a technical malfunction. The robot is nothing more than an inanimate object such as, for example, a TV set. The robot replies that he is no different from the company's other employees, and that if he does not have mental feelings, they don't have them either, because he can explain every facet of their behavior using only physical concepts. On this basis, he should have the same rights they do.

What is the point of this tale? It highlights the dilemma that the presence or absence of mental experience in an object cannot be diagnosed operationally (Squirs, 1990). In other words, there is no test capable of verifying animacy. Let us emphasize that such a test must differ fundamentally from the famous Turing test, in which a machine and a human being are given the same set of questions: if a human expert cannot distinguish their answers, this means that the machine possesses intellect. Such a test does not allow us to determine whether the machine is endowed with subjectivity.

Many of us believe that any behavior, whether that of a robot or of a human, can be explained within a natural-scientific framework. Looking into ourselves, however, we feel that reducing mental experience to physics eliminates something important. In constructing a natural-scientific picture of ourselves, we leave out our subjective mental domain. The solution may be in constructing a different picture of reality from the one we typically use. I will try to show that Plato's concept of the *ideal* may provide the basis for such a picture.

1.2. Plato's ontology

Plato, who lived two and a half thousand years ago, opposed the 'ideal' to the 'material'. This distinction became a cornerstone of Western civilization. It underlies our understanding of ourselves and of the world. This distinction also lies at the core of our understanding of morality (Sayre, 1983).

Unlike us, however, Plato believed that pure ideas exist outside of the human mind and that material objects are only dim reflection of a realm of the ideas. Following Plato, we will call an idea understood in that way the *eidos*.

Plato illustrates the realm of ideas by his famous Myth of the Cave (in *The Republic*). Let us recall this allegory. Shackled prisoners are sitting in an underground cave and cannot turn their heads.

Behind them, there is a crack through which light passes. The prisoners see the shadows of a fence and of various objects that people behind the fence are carrying high above their heads. The prisoners do not know they are seeing shadows: they think it is reality. In the allegory, the objects carried by people are analogues of pure ideas or *eide*. The shadows on the cave wall are analogues of a physical world possible to observe directly. Plato created what might be called "a path to understanding ideal entity." To progress along this path, we have to break the shackles and turn our heads toward the light.

Plato's conception did not remain the dominant one; it was replaced by that of Aristotle, Plato's disciple, who rejected his teacher's scheme concerning the real existence of ideas and stated that ideas exist only as abstractions in human minds. When we call an ideal scheme an abstraction, we follow Aristotle.

Aristotle's conception turned out to be incredibly effective. It led to the development of modern experimental and theoretical science. Despite these successes, however, there remains an unresolved problem. We have made no progress in comprehending the nature of consciousness.

1.3. Attempted solutions in physics

Many researchers have tried to find specific physical processes underlying consciousness. Special hopes were raised in connection with quantum mechanics (see Squires, 1990; Barrett, 1999; Satinover, 2001). Unlike the macro-objects studied by classical physics, micro-particles' movements are indeterminate. In the well-known experiment with two slits we can calculate the probability that a particle will hit a certain area on the screen behind the slits, but it is impossible to predict the exact place for any given particle. Since micro-particles' behavior is fundamentally indeterminate, a metaphor appeared in 1930's: the electron has free will. This metaphor is still used, in one form or another, by researchers who try to explain consciousness on the basis of quantum physics.

Another idea was proposed by Niels Bohr, one of the creators of quantum physics. In the 1920s, Louis de Broglie showed that a micro-particle ought to possess both particular and wave-like features simultaneously. The existence of such a centaur seems to contradict common sense. How can the particle be a body and a wave at the same time? To overcome this problem, Bohr formulated the principle of complementarity: we cannot observe a particle both as a body and as a wave simultaneously. In experiments of one type, we see only a particle; in experiments of another type we see only a wave. This approach underlies the Copenhagen interpretation of quantum mechanics. Bohr suggested using the principle of complementarity as a metaphor for representing mental phenomena (Bohr, 1958).

The most exciting concept in quantum mechanics is the wave function. It sets the distribution of probabilities or of densities of probabilities for various outcomes of experiments. Many researchers have tried and continue to try to understand consciousness as the special realization of a process analogous to the one whose description is given by the wave function. This route seems very attractive, but there is a serious obstacle: in the temperature range suitable for living organisms, the wave function describes only micro-processes. Macro- processes can be described by quantum mechanics only near absolute zero (Penrose, 1989). Consciousness does not seem to be a phenomenon connected to a single micro-particle, but rather with macro-processes of some kind.

In this study, we will elaborate the idea of a mechanism generating distributions of probabilities; we will connect it not only with quantum mechanical phenomena, but also with dynamic systems at their points of chaotic behavior.

1.4. The *eidos*-navigator

Let us go back to Plato's ontology and look at the world from the point of view of theoretical mechanics, attempting to find a lacuna within which consciousness may be located. Consider a moving asteroid, a material body interacting gravitationally with many other bodies. We use the term "material body" for a system in which connections between its own elements are stronger than their connections with elements external to the system. The theory of dynamic systems says that some trajectories of a body in movement may be unstable, meaning that the slightest impact is enough for the body to change its trajectory. It is assumed that the change of the trajectory under conditions of bifurcation or more complex ramifications depends on some vanishingly small force not analyzed in the theory itself (see, for example, Peitgen et al., 1992). Now, let us construct an argument making sense of this "vanishingly small impact." Let us imagine that a physical body, by its nature, is equipped with an "ideal navigator" which determines the body's behavior at the points of instability: the navigator generates a distribution of probabilities and transfers the body into a new state according to this distribution.

For the asteroid in the example above, several available trajectories, none of which contradicts the law of universal gravitation, appear at the point of instability, and we suppose that the navigator sets the distribution of probabilities and makes the *choice* of a trajectory requiring an infinitely small impact. In this argument, the material body obeys both the laws of nature and its navigator.

Suppose that the navigator participates in a physical process (not necessarily a mechanical one), if there appear special points at which the process becomes indeterminate. Suppose further that the navigator is a peculiar "factory" producing distributions of probabilities and "infinitely small" impacts directing the evolution of a material body. This "factory" is equipped with ideal physical machines consuming "ideal" energy. It is important to emphasize that the ideal machines do not consume real energy. To define the relation between the set of material bodies and the set of navigators, we assume that each body has exactly one navigator.

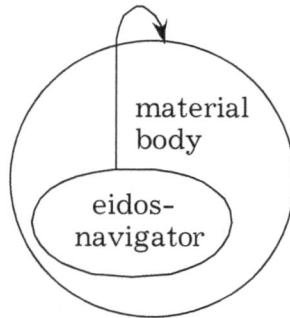

Fig. 1.4.1. A material body and its *eidos*-navigator

What is the correspondence between our scheme and Plato's? Navigators are the ideal entities, *eide*, and the real physical processes are shadows cast by the *eide* on the cave walls. We have two kinds of reality: navigators and material bodies.

Let us take the next step and assume that *a navigator's functioning is the consciousness of the material body to which it corresponds*. Therefore,

> *A material body is animate or living, if it has at least one special point (for example, a bifurcation point) at which its behavior is not determined unambiguously.*

We believe, however, that the degree of animacy depends on the complexity of the physical process. For example, a living organism comprises many complicated processes with a great number of instability points, so that we say it has a complex navigator, but a stone on the ground is not animate, since has neither "special point" nor a navigator. In the framework of our scheme, the following statements hold:

A human being has a material body.
This body has an eidos-navigator.
An ideal physical process is taking place in the navigator.
This process is human consciousness.

Therefore, human behavior is determined not only by the brain, but also by a navigator. If our assumptions are correct, consciousness possesses the structure and functional organization of a certain physical process, that is to say, consciousness is the "form of existence" of an ideal physical process.

1.5. Getting inside the navigator

How can we observe the ideal physical process going on inside the navigator? We assume that the consciousness, or, in more general terms, animacy, is an ideal physical process. If, then, we look at a mental process, we can find in it the structure and dynamics of a physical process obeying the fundamental laws of nature, for example, the first and second laws of thermodynamics. Although contemporary psychology does not have methods for detecting mental processes *per se*, in certain cases it may be possible to do.

The general scheme of our work is as follows. We take a psychological phenomenon having both mental and behavioral components, construct a corresponding mathematical model, and conduct empirical and experimental testing of the model. If the model passes the tests, we search for an ideal physical process, some aspects of which are described by the same mathematical model. This physical process (if we succeed in finding it) is a *hypothesis* about what is going on in the navigator. We have to test the hypothesis as well. The test consists of looking for other psychological phenomena different from the initial ones which can be modeled by other aspects of the ideal physical process so obtained. If such mental processes exist, it means that our hypothesis about the navigator's functioning is not rejected.

Let us emphasize that the physical process has many aspects; it is "richer" than the initial mathematical model of the psychological process. Only some aspects of the physical process are described by the initial mathematical model; there are other aspects described by other mathematical models. Thus, by constructing one mathematical model and finding a hypothetical physical process, we obtain a group of mathematical models and have to test each one by linking them with psychological processes.

Therefore, on the basis of mathematical models of certain psychological phenomena, we arrive at a hypothesis describing an ideal physical process that corresponds to them; we receive a set of other mathematical models and check whether there are psychological processes corresponding to these as well. Underlying this procedure is the expectation that the ideal process in the navigator correlates to many different psychological phenomena. We will begin our work by constructing a mathematical model of bipolar choice.

Chapter 2

The mathematical model of bipolar choice and its testing

A choice between two alternatives, one of which is linked to a positive pole and the other to a negative pole, is called *bipolar choice*. For example, a choice between truth and lie, or between the adjectives 'beautiful' and 'ugly' in characterizing an object is called *bipolar*. A bipolar choice may be complete, such as when an alternative is chosen, or incomplete, such as when a score is marked on a scale (for example, a given object is evaluated as 0.7 good and 0.3 bad). Experiments show that the mean frequency of complete choices of poles is approximately equal to the mean score on the interval [0,1]. So, we will consider a score on a scale as the probability that the subject is ready to choose the positive pole.

2.1. The mathematical model

Let us construct a mathematical model of bipolar choice (Lefebvre, 2004, 2006b). We will assume that a living organism can be represented as

$$B = \psi(z,S), \tag{2.1.1}$$

where B corresponds to the organism's behavior, z - to the influence of the environment, and S is an internal variable representing the organism's mental experience.

If we consider S as a basic variable and z as a parameter, then (2.1.1) can be written as

$$B = \Phi_z(S). \tag{2.1.2}$$

For bipolar choice we must connect B with the probability of choosing one of the poles, for example, the positive pole, and z with probability of the environment's pressure toward the positive pole at the moment of making choice, x_1. Then we can write (2.1.2) as

$$X(S) = F_{x_1}(S) . \tag{2.1.3}$$

We assume that $X(S)$ is a differentiable function, where $0<X(S)<1$, $0< x_1 <1$, $S \geq 0$ and

$$X(0) = x_1. \tag{2.1.4}$$

Condition (2.1.4) means that at $S=0$ the probability of choosing the positive pole is equal to the environment's pressure toward the positive pole at the moment of making the choice.

Suppose that it takes a certain time to make the choice. If the value of S is constant, the probability of choosing the positive pole is X_S. However, if S increases by some small value ΔS during this interval, then the choice procedure becomes different: we call this the *axiom of repeated choice*.

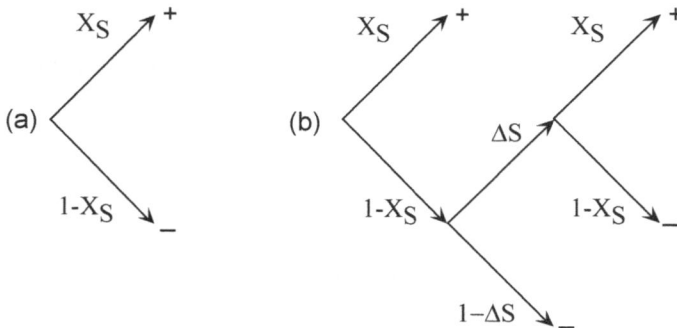

Fig. 2.1.1. Choice trees: (*a*) for the value of the internal variable equal to S; (*b*) for the value of the internal variable equal to $S+\Delta S$

First, the subject makes a choice with probability X_S of choosing the positive pole. If the positive pole is chosen, the subject executes his choice. If the negative pole is chosen, the subject cancels his choice with the small probability ΔS and repeats the procedure of making choice. The result of the repeated choice is then executed. The two trees in Fig.2.1.1 depict the axiom of repeated choice for S and $S+\Delta S$.

In accordance with tree (b),

$$X_{S+\Delta S} = X_S + (1 - X_S)\Delta S X_S \ . \tag{2.1.5}$$

Now we will search for a differentiable function $X(S)$ represented as

$$X(S + \Delta S) = X(S) + (1 - X(S))\Delta S X(S) + o(\Delta S) \ .$$

After transformations and passage to the limit at $\Delta S \to 0$ we obtain the differential equation

$$\frac{dX(S)}{dS} = (1 - X(S))X(S) \ . \tag{2.1.6}$$

Solving it using (2.1.4) as initial condition, we obtain the logistic function

$$X(S) = \frac{x_1}{x_1 + (1 - x_1)e^{-S}} \ . \tag{2.1.7}$$

Henceforth we will write this expression as

$$X_1 = \frac{x_1}{x_1 + x_2 - x_1 x_2} \ , \tag{2.1.8}$$

where

$$x_2 = e^{-S} \ . \tag{2.1.9}$$

The value of the internal variable S is interpreted as the level of importance, for the subject, of choosing the positive pole. With constant x_1, the greater S, the greater X_1.

Let us examine (2.1.9). We interpret the value of x_2 as the subject's estimation of the probability with which the world inclines him toward choice of the positive pole. It is important to emphasize the difference between x_1 and x_2: x_1 is the world's objective pressure to choose the positive pole, and x_2 is a subjective evaluation of that pressure. Also, x_1 is a *local* characteristic of the environment at the *given moment*, and x_2 is a *global* characteristic based on the subject's experience in the presence and past. It is a measure of the world's positivity for the subject.

We assume also that the subject has an image of the self, X_2, which is the subject's estimation of the probability of his choosing the positive pole. While X_1 is an objective characteristic that belongs to the *real* world, X_2 is a subjective characteristic belonging to the mental world.

The subject with image of the self can be represented as a box with input x_1 and output X_1, into which another box has been inserted with input x_2 and output X_2 (Fig. 2.1.2).

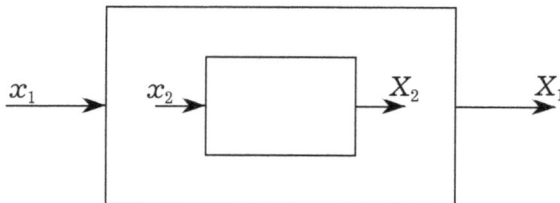

Fig. 2.1.2. Subject with image of the self

We assume that

$$\frac{X_1}{x_1} = \frac{X_2}{x_2}.$$

(2.1.10)

In other words, the subject's image of the self is similar to the subject, in the sense that they equally amplify an input signal. It follows from (2.1.8) and (2.1.10) that

$$X_2 = \frac{x_2}{x_1 + x_2 - x_1 x_2}. \qquad (2.1.11)$$

Expressions (2.1.8) and (2.1.11) play an important role in our further considerations.

The concept of 'image' can be generalized by assuming that the image of the self may have an image of the self and that the latter also has an image of the self etc. We assume that expression (2.1.10) holds for each image and its image of the self. Fig. 2.1.3 depicts the subject with a chain of images:

Fig. 2.1.3. Subject with chain of images: 1- subject, 2 - subject's image of the self, 3 - image of the self of the image of the self, etc.

The arrows depict the relation "knows"; the statement "1 knows 2,who knows 3,... who knows n" corresponds to the entire chain.

Let us introduce the relation "is aware of," equivalent to double relation "knows": if A "knows" B, and B "knows" C, then A "is aware of" C. We depict this relation with a curved arrow.

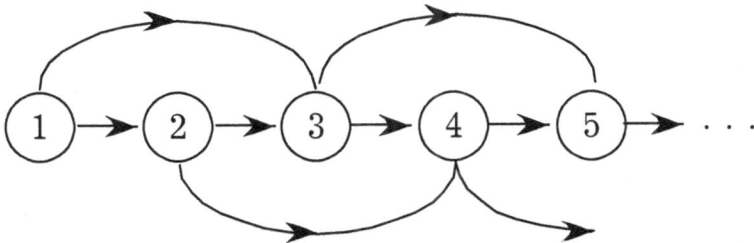

Fig. 2.1.4. Subject with chain of images and relations: "knows" and "is aware of"

Subject 1 is aware of image 3, image 2 is aware of image 4, image 3 is aware of image 5, etc. We call a subject or image from which an arrow originates an *original*, and one to which an arrow points a *copy*. We say that a copy is equivalent to the original if they are characterized by the same pair x, X, where x is the probability of pressure toward the positive pole, and X is the probability of choosing the positive pole. Let us introduce a postulate of equivalency:

In awareness, the original and the copy are equivalent.

It follows from this postulate that elements with odd numbers are characterized by a pair x_1, X_1, and those with even numbers by a pair x_2, X_2, given by the expressions (2.1.8) and (2.1.11). This correspondence is depicted in Fig. 2.1.5.

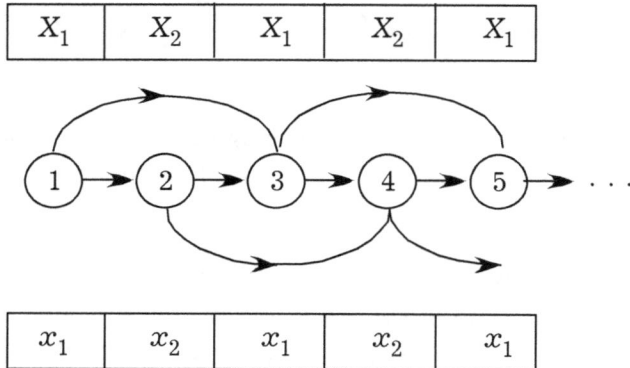

X_1	X_2	X_1	X_2	X_1

x_1	x_2	x_1	x_2	x_1

Fig. 2.1.5. Values of pairs corresponding to the subject
and his images of the self

The scheme in Fig. 2.1.5 corresponds to the mathematical model of bipolar choice; it generates two pairs of distributions:

$$(X_1, 1 - X_1), (x_1, 1 - x_1) \quad \text{and} \quad (X_2, 1 - X_2), (x_2, 1 - x_2).$$

After conducting the necessary empirical examination, we will demonstrate that these distributions can be formally represented as the product of work by a chain of heat engines which can perform functions different from those in "real life," namely, they can generate probability distributions. Therefore, unlike quantum mechanics, where the probability distribution is found with a Fourier series, in our model the probability distribution is found through representing the subject's mental domain as a chain of heat engines (Chapter 3).

2.2. Categorizing stimuli

Categorization is evaluation of a stimulus over a set of intensity levels. For example:

very weak   weak   moderate   strong   very strong

Usually, numbers are used instead of verbal evaluations:

1 2 3 4 5

The procedure is as follows. First, the subject is presented with the weakest and strongest stimuli to be used in a particular experiment, for example, two bars of 5 cm and 105 cm; after that, all other stimuli are presented one by one. The subject's task is to assign each stimulus to a certain category. The data obtained allow construction of a graph linking the physical intensity of the stimuli with their categorical estimation. Researchers were surprised by the initial results. They expected to find a linear relation between physical measures and categorical estimations of such qualities as length, area, and duration. Instead, they found curved graphs.

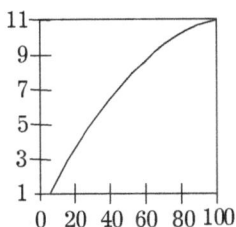

Fig. 2.2.1. Estimation of the length of metal bars.
Length in centimeters is mapped onto the X-axis, and categorical estimation onto the Y-axis (data by Stevens & Galanter, 1957)

For example, the graph in Fig. 2.2.1 shows the results of an experiment on length estimation.

The experiments that followed demonstrated that the graph's curvature depends on the distribution of the intensity of stimuli. If, in a given experiment, there are more weak stimuli, the curvature increases; if there are more strong stimuli, the curvature decreases (Fig. 2.2.2).

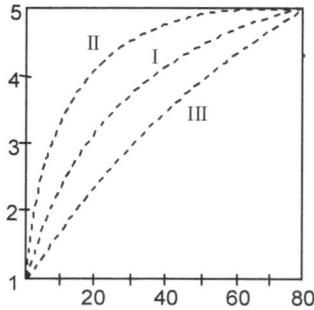

Fig. 2.2.2. Distribution of stimuli: I - weak and strong stimuli presented equally; II - weak stimuli presented more often; III - strong stimuli presented more often (Stivens & Galanter, 1957)

A problem of categorical graph curvature has puzzled researchers for more than sixty years, but no truly convincing explanation has been found. We will reexamine this problem from the point of view of the formal model of bipolar choice (Fig. 2.1.2). Let us introduce the following parallels:

x_1 is the normalized physical intensity of a presented stimulus,

x_2 is the mean normalized intensity of all stimuli presented to the subject,

X_1 is the normalized categorical estimation of a given stimulus, in the following called simply "categorical estimation."

In the mathematical scheme of bipolar choice, the process of categorization is given by the function

$$X_1 = \frac{x_1}{x_1 + x_2 - x_1 x_2} \ . \tag{2.2.1}$$

The choice is not complete. Variable x_1 represents perception, x_2 represents memory, and X_1 is the outcome of the stimulus estimation.

Consider the experiment with categorization of bar length. First, two bars are presented: the shortest (5 cm) and the longest (105 cm). The former acquires the role of a negative pole, and the latter that of a positive pole. The stimulus intensity is the length of the bar. From the semantic point of view, the bar's length equates to its endowment with positive value, because in the construct *long-short*, the adjective *long* plays the role of the positive pole, and *short* that of the negative pole.

At the level of perception, the subject's cognitive system determines the value

$$x_1 = \frac{\psi - \psi_{\min}}{\psi_{\max} - \psi_{\min}}, \qquad (2.2.2)$$

where ψ is the physical length of the bar currently presented, $\psi_{\max}$ that of the longest bar, and $\psi_{\min}$ that of the shortest bar in the series. The value of x_1 is the pressure toward the positive pole at the level of perception. Then, the subject's cognitive system finds the mean value of the intensity of the stimuli previously presented:

$$x_2 = \frac{x_1^{(1)} + x_1^{(2)} + \ldots + x_1^{(n)}}{n}. \qquad (2.2.3)$$

The value of x_2 represents the class of stimuli shown to the subject. It is the level of the world's positivity in the context of a given series of estimations. Obtaining the values for x_1 and x_2 from (2.2.2) and (2.2.3) and substituting them into (2.2.1), we find the categorical estimation for a given bar.

Let us consider a set of experiments with different values of x_2. For each x_2, there is a curve corresponding to function X_1. The smaller the value of x_2, the more convex the curve (see Fig. 2.2.3).

When the intensities of stimuli are shifted toward the weakest stimulus, the mean intensity decreases, and *for that reason* the curve representing the data is more convex. When the intensities of stimuli are shifted toward the strongest one, the mean intensity increases, and *for that reason* the curve representing those data is less convex.

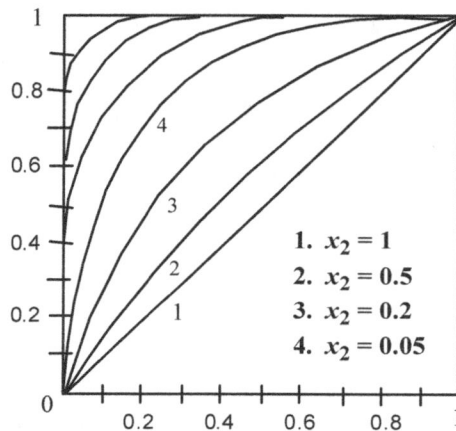

Fig. 2.2.3. Family of curves corresponding to function X_1
for different values of x_2

Therefore, our model explains the phenomena appearing in experiments with categorization.

2.3. The Golden section phenomenon in choice and categorization

In the mid 1950s, George Kelly (1955) proposed the hypothesis that, in each person's cognitive sphere, there exists a unique system of bipolar category constructs such as *active-passive, sharp-blunt, strong-weak,* etc. This system allows an individual to have a multivariate view of other people's qualities and personalities. According to Kelly, the constructs inherent to an individual are bipolar, with positive and negative poles that are used with equal frequency if the person evaluates a large number of other people. Kelly's disciples, however, Adams-Webber and Benjafield, found (1973), that equal estimations are very rare and that the frequency of choosing the positive pole is 0.62. They hypothesized that the theoretical value of the frequency of using the positive pole in evaluating others is the famous golden section $(\sqrt{5} - 1) / 2 = 0.618...$, a ratio associated with beauty and attractiveness from antiquity to the present (Benjafield, & Adams-Webber, 1976).

Having analyzed numerous experiments with bipolar estimations, we found that the golden section ratio is indeed present both in evaluating persons and in evaluating inanimate objects (Lefebvre, 1985, 1987, 1992). It appears precisely in those cases when the subjects do not have operational means for isolating the quality they must evaluate.

For example, let a subject is given a picture of a flower and a scale for evaluation of the flower's beauty (Fig. 2.3.1).

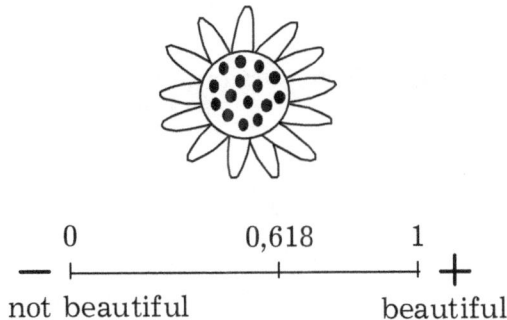

Fig. 2.3.1. Stimuli categorization

We will try to understand the structure of the computation performed by the subject's cognitive system in such cases. First, it polarizes the scale: "beautiful" becomes the positive pole, and "not beautiful" becomes the negative pole. Second, the subject's cognitive system provides values for the variables x_1 and x_2. Since beauty does not have direct physical intensity, we may suppose that x_1 takes on a neutral value

$$x_1 = \frac{1}{2}. \tag{2.3.1}$$

For the same reason, we assume that the value of x_2 is not a physical measure. It is equal to the score of the flower's beauty generated by the subject:

$$x_2 = X_1. \tag{2.3.2}$$

The flower, for the subject, is both the world and, at the same time the object of evaluation. By substituting the values of (2.3.1) and (2.3.2) into the expression

$$X_1 = \frac{x_1}{x_1 + (1 - x_1)x_2}, \tag{2.3.3}$$

we obtain the following equation

$$X_1^2 + X_1 - 1 = 0. \tag{2.3.4}$$

Its positive root is the golden section:

$$X_1 = \frac{\sqrt{5} - 1}{2} = 0.618\ldots. \tag{2.3.5}$$

This is the probability with which the subject is ready to choose the positive pole in evaluating the flower.

We decided to test this prediction experimentally and empirically.

Experiment with pinto beans

The experimental material consisted of fifty small transparent envelopes, each containing two pinto beans. The beans were chosen to have as few differences as possible. The subject's task was to pick up the envelopes one by one, evaluate a pair as good or bad, and drop the envelope into one of two boxes marked "+" and "-". The ratio of positive estimations was 0.611 (V.D. Lefebvre, 1990). This experiment supports the hypothesis that the theoretical value of positive evaluation in the absence of an operational criterion is equal to the golden section, i.e., to the predicted value. When the experiment with pinto beans was replicated, the ratio of positive estimation was equal to 0.64, insignificantly different from the golden section value (Anderson and Grice, 2009).

Moral judgment

Independently of our research, McGraw (1985) conducted the following experiment. Each subject was given a scenario describing either a good action or a bad action. For example, someone found a wallet and returned it to its owner; a blind girl dropped money and someone helped her to collect it: these are good actions. Or, the wallet was not returned to the owner; the blind girl was not helped to collect her money: these are bad actions. The task was to predict the percentage of students who would act in the way described by a given scenario. The mean estimation by those who received good scenarios was 62%, as opposed to 39% by those who received bad ones.

We can explain these results as follows. For all subjects the positive pole is "100% good students" and the negative pole is "0% good students". Thus, for all subjects the mean score is close to the golden section value, 62% and 61%. The subjects who received the bad scenarios marked the percentage of bad students, 39%, as the instructions required. In this experiment, $x_1 = \frac{1}{2}$, because pressures toward both poles were the same, and $x_2 = X_1$, i.e., the subjects did not have previous experience of similar experiments. Under these conditions, equation (2.3.3) turns into (2.3.4), whose positive root is the golden section value.

Mere exposure

The phenomenon of the golden section manifests itself in the choice between two objects that are barely distinguishable, provided that one of them is associated with a positive pole and the other with a negative pole. To polarize the objects, one of them is shown beforehand, making it more likely to become the positive pole. To avoid determining in advance the value of x_2 (which is related to previous conscious experience), the exposure must be subliminal.

Let us look at one such experiment conducted without any relationship to our model (Kunst-Wilson & Zajonc, 1980). Twenty improper octagons were used as the experimental material. For each subject, the twenty octagons were divided into two groups of ten. In the first part of the experiment, the subject was shown ten octagons from one of the groups five times each with an exposure time of one millisecond. This short exposure time does not allow the subject to perceive the object consciously. In the second part of the experiment, the subject was presented with a series of pairs of octagons, each consisting of one octagon previously shown and one new. The experimenter did not inform the subjects that the previously shown figures were among the octagons presented in the second part of the experiment. The task was to choose, in each pair, the octagon that the subject "likes more." The subjects chose the one presented in the first part of the experiment with a frequency of 0.60.

Below are frequencies of choosing the alternatives presented in advance, obtained from other similar experiments:

Seamon et al. (1983)	0.61
Mandler et al. (1987)	0.62
Bonano et al. (1986)	0.66; 0.63; 0.62; 0.61; 0.63; 0.62

We see that the frequencies are grouped around 0.62 (a fact that the experimenters apparently did not notice). The preliminary exposure of one of the alternatives polarized the pairs in which it would subsequently appear; the one that had been shown previously became the positive pole, and the other, the new one, the negative pole. Attractiveness is not here a measurable quality, so that $x_1 = \frac{1}{2}$.

The stimuli used in the experiment had not previously been evaluated by the subjects, so that the subjects had no history of evaluating them. This explains the appearance of the golden section ratio.

Medians in referenda

The mathematical model of bipolar choice predicts a hitherto unremarked phenomenon involving the results of referenda. For example, the following draft bill is offered to voters:

Proposed, to reduce the number of optional subjects in public schools, in order to better teach required subjects, on the one hand, and to facilitate balancing the state budget, on the other.

YES _____ NO _____

Fig. 2.3.2. The draft bill

Let us assume that the formulation of this question in a given social context polarizes the alternatives "yes" and "no" in such way that for the majority of voters one of them becomes the positive pole and the other the negative pole. For example, let's say that shortly before the referendum there was an active campaign in favor of broad education in public schools. Thus, the role of positive pole is played by the alternative "no".

At the moment of casting a ballot, the words "yes" and "no" are equal; thus, $x_1 = \frac{1}{2}$. An ordinary voter has no in-depth experience in educational policy; thus $x_2 = X_1$, and the phenomenon of the golden section should appear. Therefore, under ideal conditions, people should choose the positive pole with a frequency of 0.62. In reality, however, there are voters who think about the issue and make their decision in advance, not at the moment of casting their ballot. We may think there are not too many of them, but there are enough to decrease or increase the frequency of the positive choice relative to 0.62. Further, we presume that:

- decreases or increases in the frequency of choosing the positive pole in relation to 0.62 happen equally often on the set of all referenda and

- in each referendum, the frequency of choosing the positive pole does not drop below 0.5 .

Under these assumptions, *the median distribution* of winning poles must be equal to the golden section value, i.e., the number of polls winning with lower than 62% votes must be equal to the number of polls winning with higher than 62% votes.

We have analyzed all the referenda in California from 1884 to 1990 (Eu, 1983a,b; 1985a,b; 1987; 1989a,b,c).

Figure 2.3.3 (next page) represents the distribution for referenda analyzed. Its median is equal to 62%.

A similar analysis was conducted for referenda in Switzerland 1886-1978 (Butler & Ranney, 1978). The median found was 63%. We also analyzed Oregon referenda 1904-1914 (Barnett, 1915) and found the median equal to 62%. Finally, we used the results of US referenda for 1977-1988 (Ranney, 1978; 1981; 1983; 1985; 1987; 1989). The median again was equal to 62%.

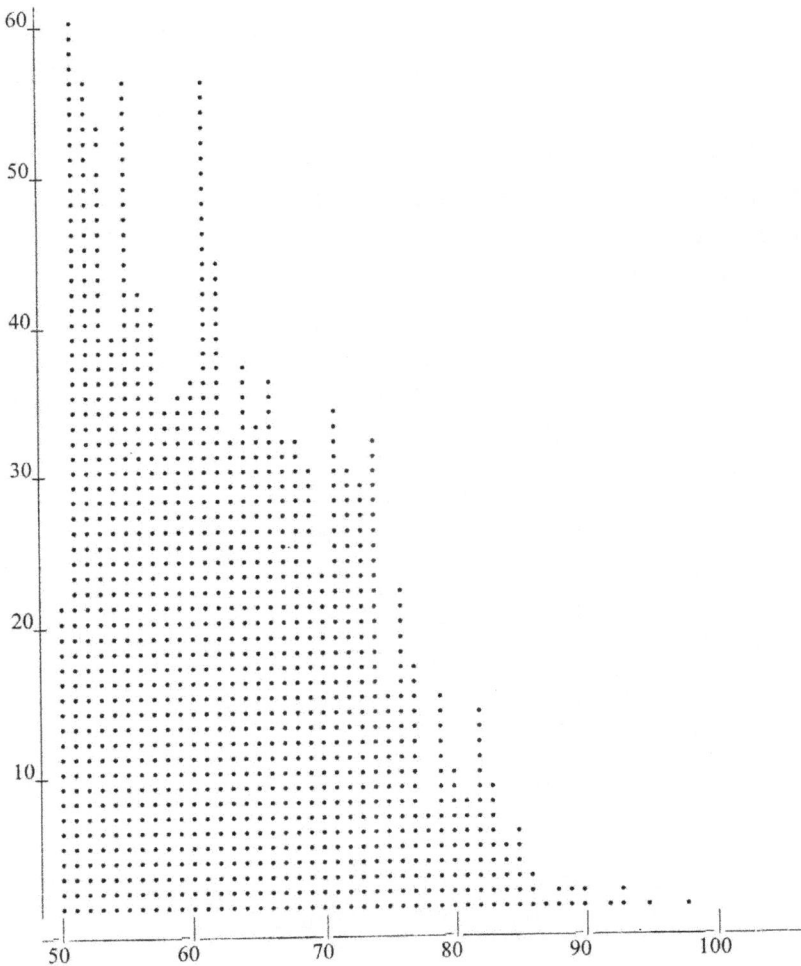

Fig. 2.3.3. Distribution of winning polls in California referenda 1884-1990. Each dot represents one referendum. The horizontal axis is the percentage scale 50-100%. Each column of dots shows the number of polls won with the given percentage of votes.

The two-humped graph

Let us consider an experiment by Poulton and Simmonds (1985). The subjects' task was to evaluate the degree of lightness of a sample of gray paper situated between black and white papers. The tone of the gray sample was picked in such a way that, on a psychological scale, it was exactly in the middle between the tones of the black and white samples. Each subject had to mark his evaluation of the gray sample's lightness on a hundred-millimeter scale, one end of which corresponded to black and the other to white. Only the very first touch of the subject's pencil was taken into account. The result of the experiment is given in Fig. 2.3.4; it is a two-humped graph with a dip in the middle.

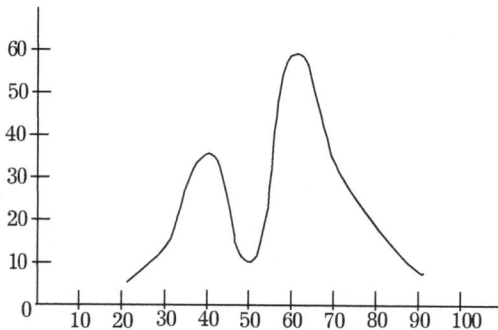

Fig. 2.3.4. A two-humped graph obtained in the experiment
with lightness estimation

Psychophysics cannot explain this phenomenon, but we can explain it using the mathematical model of bipolar choice. Suppose that one portion of the subjects took the white sample as the positive pole, and the other part took the black sample as positive. Thus, the first group was evaluating the degree of whiteness of the gray sample, and the second group was evaluating the degree of its blackness. Since the pressures toward the two poles were equal,

$x_1 = \frac{1}{2}$. The subjects did not have experience in similar estimations (only the first touch was counted), so $x_2 = X_1$. Therefore, in both groups the estimation was equal to the golden section value. The peaks of the humps are located at 62 mm from the white end and 62 mm from the black end of the scale.

2.4. Altruism

It has been known for centuries that people who have committed some act they repent of are more inclined toward altruism. In the twentieth century, this opinion has been supported experimentally. Moreover, a related phenomenon was discovered. It turned out that people become more inclined to altruism not only when they feel guilty of something, but also as a result of observing the world's unfairness.

Here is one of the experiments in question (Regan, 1971). The student-subjects were asked to give a slight electric shock to a rat. Unnoticeably for the subject, the experimenter increased the amperage so that the rat jumped. The subjects were divided into three groups. In the first group, the amperage was not changed. In the second group, the amperage was increased but the subjects were told that they were not at fault, it was a short circuit. In the third group, after the rat jumped, the subjects were told that they had made a mistake, and that the experiment would have to be stopped.

Subsequently, each subject was involved in a situation in which he had the opportunity to donate small amount of money to a summer student project. The subjects from the second and third groups donated money significantly more often than those from the first group.

Let us consider this experiment from the point of view of the mathematical model of bipolar choice.
1. There are two alternatives: to donate money or not to donate money.

2. Variable x_1 is the pressure on the subject at the moment of the request for a donation; variable x_2 is the level of the world's positivity at that moment; X_1 is the probability with which the subject will choose the positive pole; X_2 is the subject's estimation of his readiness to choose the positive pole.

3. These variables are connected with the following expressions:

$$X_1 = \frac{x_1}{x_1 + x_2 - x_1 x_2} \, , \qquad\qquad (2.4.1)$$

$$X_2 = \frac{x_2}{x_1 + x_2 - x_1 x_2} \, . \qquad\qquad (2.4.2)$$

In the first group, the values of x_1 and x_2 are not changed in the course of the experiment. In the second group, the value of x_2 changes after the subject sees the rat suffering; it decreases, because the world has become less positive. It follows from (2.4.1) that, with the value of x_1 remaining constant and x_2 decreasing, the value of X_1 increases monotonically. Thus, the subjects should become more altruistic, as was found in the experiment.

In the third group, the experimenter influenced the subject's image of the self. It follows from (2.4.1) and (2.4.2) that

$$X_1 = 1 - (1 - x_1)X_2 \, . \qquad\qquad (2.4.3)$$

With a constant value of x_1 we can consider X_2 an independent variable (such that variable x_2 becomes dependent). The feeling of guilt means decreasing the degree of positivity of one's image of the self, i.e., decreasing the value of X_2. It follows from (2.4.3) that with decreasing X_2, X_1 increases, i.e., the subject demonstrates greater altruism.

We see that the mathematical model of bipolar choice explains the increase of altruism resulting both from observing the world's unfairness and from feeling one's own guilt.

2.5. Bipolar choice in birds and animals

Almost a half-century ago, Richard Herrnstein published the results of his experiments with birds (1970). A pigeon was placed in a Skinner chamber with two feeders; by pecking the feeder, the bird received a small ration of grain. The feeders were controlled by independent reinforcement programs that the experimenter could vary. Under the influence of these programs, one might suppose that the pigeon's behavior could be shaped in different ways, but that was not the outcome. The pigeons used one particular strategy: their frequency of responses was approximately proportional to the frequency of reinforcement. This finding was called the Matching Law. Subsequent experiments showed systematic deviations from proportionality: birds were pecking one of the feeders more often than the Matching Law predicted. Moreover, it was found in many experiments that pigeons and rats turn to the "poorer" feeder more often than proportionality would require. Various other correlations were suggested, and, finally, William Baum (Baum et al., 1999) offered a new formula producing good predictions:

$$\frac{B_2}{B_1} = b\frac{r_2}{r_1}, \qquad (2.5.1)$$

where B_1 is the frequency of going to one feeder; B_2 is the frequency of going to the other feeder; r_1 is the frequency of reinforcements given in the first feeder, and r_2 the frequency given in the second. We can always choose the feeders' numeration such that $b \leq 1$ holds.

It should be noted that only recently has the model of bipolar choice begun to be used in research on animal choice (Lefebvre & Sanabria, 2008). Let us write (2.5.1) as

$$\frac{1 - B_1}{B_1} = b\frac{1 - r_1}{r_1} \qquad (2.5.2)$$

and then rewrite it as

$$B_1 = \frac{r_1}{r_1 + b - r_1 b}.$$

(2.5.3)

We see that (2.5.3) coincides with the expression

$$X_1 = \frac{x_1}{x_1 + x_2 - x_1 x_2},$$

(2.5.4)

which shows the probability of choosing the positive pole in the mathematical model of bipolar choice. This fact suggests that animals, as well as humans, use binary valuations *'good-bad'*.

This problem is described in more detail in Appendix III.

Chapter 3.
The physical model and its testing

In the previous chapter we presented a number of empirical and experimental arguments supporting the contention that the mathematical model of bipolar choice, as set by the correlations

$$X_1 = \frac{x_1}{x_1 + x_2 - x_1 x_2}, \tag{3.1}$$

$$X_2 = \frac{x_2}{x_1 + x_2 - x_1 x_2}, \tag{3.2}$$

manifests itself in human and animal choice. We hypothesized that consciousness is an ideal physical process. In our further considerations, we will assume that expressions (3.1) and (3.2) are related not only to the behavior of living creatures but to their mental experience as well. In this chapter, we will search for ideal physical processes that are described by those correlations.

3.1. Heat engine

Consider an abstract heat engine (Fig. 3.1.1).

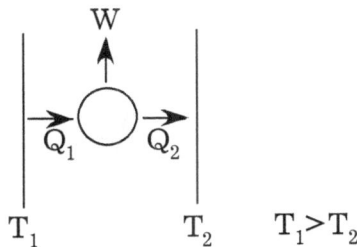

Fig. 3.1.1. Abstract heat engine

The engine takes heat Q_1 from a hot reservoir with temperature T_1, produces work W, and gives heat Q_2 to a cold reservoir with temperature T_2. The functioning of the engine obeys the first and second laws of thermodynamics:

I. The law of conservation of energy:

$$Q_1 = Q_2 + W \; ; \tag{3.1.1}$$

II. The law of undiminished entropy:

$$H_2 \geq H_1 , \tag{3.1.2}$$

where

$$H_1 = \frac{Q_1}{T_1} \tag{3.1.3}$$

is the decrease of entropy resulting from taking heat Q_1 from the hot reservoir, and

$$H_2 = \frac{Q_2}{T_2} \tag{3.1.4}$$

is the increase of entropy resulting from returning heat Q_2 to the cool reservoir.

The following values will be used in our discussion.
Entropy change:

$$\Delta H = \frac{Q_2}{T_2} - \frac{Q_1}{T_1} \; . \tag{3.1.5}$$

Efficiency:

$$\rho_1 = \frac{Q_1 - Q_2}{Q_1} \; . \tag{3.1.6}$$

Maximal efficiency:

$$\rho_0 = \frac{T_1 - T_2}{T_1} \; . \tag{3.1.7}$$

Relative efficiency:

$$\omega_1 = \frac{\rho_1}{\rho_0} . \tag{3.1.8}$$

Lost available work:

$$\Delta W_1 = T_2 \left(\frac{Q_2}{T_2} - \frac{Q_1}{T_1} \right) . \tag{3.1.9}$$

The main concepts of thermodynamics are described in Appendix I.

Compensation process

For reversible heat engines, the change of entropy is equal to zero, and the lost available work is also equal to zero. For non-reversible heat engines, the lost available work is greater than zero.

Let us include one more engine between the hot reservoirs 2 and 3 with temperatures T_2 and T_3, where

$$\frac{T_2}{T_3} = \frac{T_1}{T_2} . \tag{3.1.10}$$

Engine 2 takes from reservoir 2 heat equal to that given to it by engine 1 and produces work equal to the lost available work of engine 1. In this way, engine 2 compensates for the "imperfection" of engine 1 (Fig. 3.1.2).

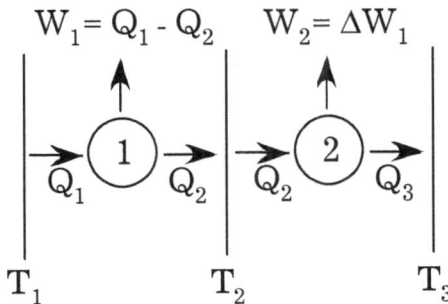

Fig. 3.1.2. Engine 2 compensates for the imperfection of engine 1

Engine 2's efficiency is

$$\rho_2 = \frac{\Delta W_1}{Q_2} = \frac{T_2\left(\dfrac{Q_2}{T_2} - \dfrac{Q_1}{T_1}\right)}{Q_2} \; , \tag{3.1.11}$$

and its relative efficiency is

$$\omega_2 = \frac{\rho_2}{\rho_0} . \tag{3.1.12}$$

It follows from (3.1.6) and (3.1.11) that

$$\frac{T_1 - T_2}{T_1} = \rho_1 + \rho_2 - \rho_1\rho_2 \tag{3.1.13}$$

and

$$\omega_1 = \frac{\rho_1}{\rho_1 + \rho_2 - \rho_1\rho_2} , \tag{3.1.14}$$

$$\omega_2 = \frac{\rho_2}{\rho_1 + \rho_2 - \rho_1\rho_2} . \tag{3.1.15}$$

Expressions (3.1.14) and (3.1.15) coincide with (3.1) and (3.2), which correspond to the mathematical model of bipolar choice if we assume that

$$\begin{aligned}
\rho_1 &= x_1, \\
\rho_2 &= x_2, \\
\omega_1 &= X_1, \\
\omega_2 &= X_2.
\end{aligned} \tag{3.1.16}$$

We can now write:

$$\frac{T_1 - T_2}{T_1} = x_1 + x_2 - x_1 x_2 \; , \tag{3.1.17}$$

$$\omega_1 = \frac{x_1}{x_1 + x_2 - x_1 x_2}, \tag{3.1.18}$$

$$\omega_2 = \frac{x_2}{x_1 + x_2 - x_1 x_2}. \tag{3.1.19}$$

The first engine stands for the subject, and the second for the subject's image of the self. Thus, we have found an ideal physical process related to the mental process of bipolar choice (Lefebvre, 1997, 2006a).

— · —

The mathematical model of bipolar choice contains the following function:

$$x_2 = e^{-S}, \tag{i}$$

where x_2 is the value of the world's positivity in the subject's mental world and S is the internal variable representing the degree of importance, for the subject, of choosing the positive pole.

Let us find an analogue for variable S in the heat engine model. Since $x_2 = \rho_2$, we can use (3.1.11) and write:

$$x_2 = \frac{\dfrac{Q_2}{T_2} - \dfrac{Q_1}{T_1}}{\dfrac{Q_2}{T_2}}. \tag{ii}$$

The right side of this expression is the normalized value of the first engine's change of entropy; we designate it E. The x_2 can be written as

$$x_2 = e^{-\ln\frac{1}{E}}. \tag{iii}$$

By comparing (iii) and (i) we see that

$$S = \ln\frac{1}{E}.\tag{iv}$$

Thus, in the heat model, the internal variable S is the logarithm of the inverse normalized value of entropy change. The lesser E, the greater S.

Heat unit

Generally speaking, the second engine in Fig. 3.1.2 is also non-reversible. It also loses available work, and we can put a third engine after it, and then a fourth one, and so forth.

Consider a sequence of heat reservoirs whose temperatures decrease in geometrical progression. Between every two reservoirs we place an engine, so that every subsequent engine takes from its hot reservoir the heat returned to it by the preceding engine and produces the work equal to the available work lost by the preceding engine. Such a sequence is shown in Fig. 3.1.3.

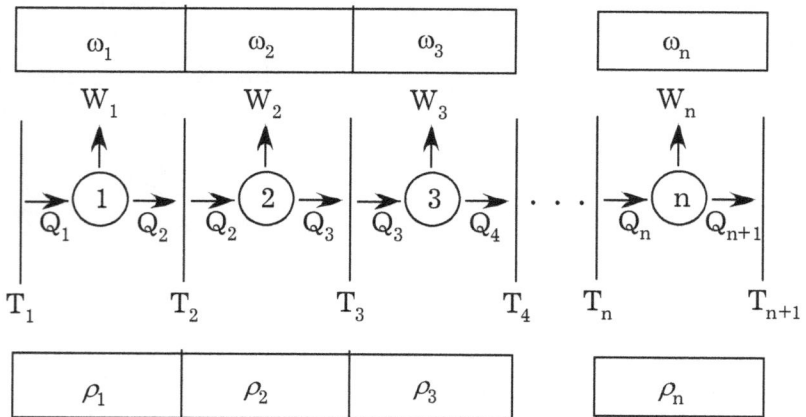

Fig. 3.1.3. Heat unit with header and footer

In addition, this unit is supplied with a header and footer; for each engine there is one corresponding cell on both header and footer. On the lower cells, the efficiencies are printed, and on the upper cells the relative efficiencies.

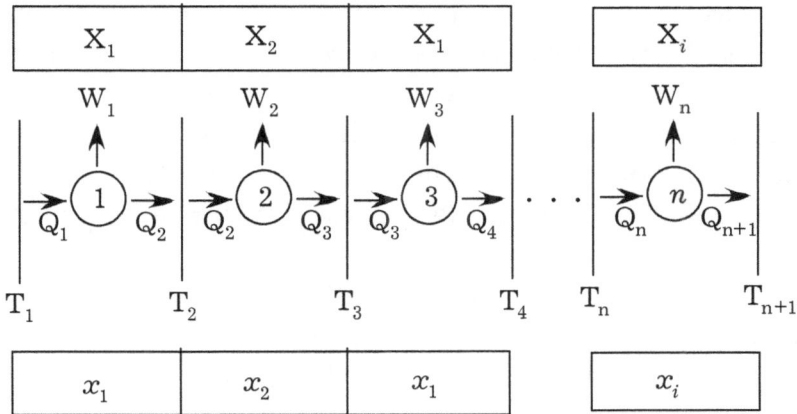

| X_1 | X_2 | X_1 | | X_i |

W_1 W_2 W_3 W_n

$\xrightarrow{Q_1}(1)\xrightarrow{Q_2}\;\xrightarrow{Q_2}(2)\xrightarrow{Q_3}\;\xrightarrow{Q_3}(3)\xrightarrow{Q_4}\cdots\xrightarrow{Q_n}(n)\xrightarrow{Q_{n+1}}$

T_1 T_2 T_3 T_4 T_n T_{n+1}

| x_1 | x_2 | x_1 | | x_i |

Fig. 3.1.4. Unit generating the mathematical model
of bipolar choice (for odd n, $i = 1$; for even n, $i = 2$)

The following statement holds:
Sequences ρ_n and ω_n are periodic: for engines with odd numbers, $\rho_n=\rho_1$ and $\omega_n=\omega_1$, and for engines with even numbers, $\rho_n=\rho_2$ and $\omega_n=\omega_2$ (Lefebvre, 1997).

Using (3.1.16), we can depict the sequence of engines as shown in Fig. 3.1.4, where X_1 and X_2 are linked with x_1 and x_2 by expressions (3.1) and (3.2).

We see that the unit generates the scheme given in Fig. 2.1.5, which corresponds to the mathematical model of bipolar choice, and that the alternation of variables on the headers and footers happens automatically without recourse to the postulate of equivalency (see Chapter 2).

Therefore, the unit in Fig. 3.1.4 has a psychological interpretation. It depicts the subject who has an image of the self, which, in turn, has an image of the self, which also has an image of the self, and so forth. The analogy between the heat unit and the hierarchy of images may be established by two ways. In the first way, which we call *descending reflexion*, the first, leftmost engine represents the subject; the second engine is the subject's image of the self; the third engine is the image of the self's image of the self, etc.

In the second way, called *ascending reflexion*, the engine with number n, the rightmost engine, represents the subject; the engine with number n-1 is the subject's image of the self; the engine n-2 is the image of the self's image of the self, etc. Under descending reflexion, the images go in the order of decreasing temperature; under ascending reflexion, the images go in the order of increasing temperature[1].

The first step has been taken. We have found a formal connection between the mathematical model of bipolar choice and an ideal physical process which obeys two fundamental laws: the first and second laws of thermodynamics. This process cannot be situated in the physical brain of human beings or animals. It is possible that we may shed light on the work of the *eidos*-navigator (see Conclusion). Now, we have a hypothesis to test.

3.2. Two laws of psychophysics

Our task is to find out if there are psychological phenomena which cannot be explained by the mathematical model of bipolar choice but can be explained by the physical model constructed in the previous section. In this section, we will demonstrate that the main laws of classical psychophysics - the Weber-Fechner law (Fechner, 1860) and the Stevens law (1975) - can be deduced from our physical model. In addition, with the help of this model we will deduce the set of harmonical intervals in music.

The research into mathematical correlations between the physical characteristics of stimuli and their subjective perception began in the nineteenth century. Gustav Fechner, relying on Weber's experiments and ideas, demonstrated that the ratio of intensities of psychological perception is proportional to the ratio of logarithms of physical intensities. For example, if the physical weights of three

---

[1] This definition of descending and ascending reflexion seems more natural than the one given in my book *Cosmic Subject*.

pieces are related as 2:8:16, the intensities of their psychological feelings are related as 1:3:4, that is, the ratio between physical stimuli diminishes in perception: the third weight is eight times heavier than the first one, but a person feels that it is only four times heavier. The Weber-Fechner law can be written as follows:

$$\psi_F = c \log \frac{\varphi}{\varphi_0}, \qquad (3.2.1)$$

where φ is the physical intensity of a stimulus, φ_0 the threshold value at which the stimulus begins to be perceived, ψ_F the psychological evaluation of the stimulus, and c the coefficient of proportionality.

Fechner's contemporaries were greatly impressed by his discovery of the logarithmic law. It seemed that a law linking the physical world with the human mental domain had finally been found. Some people believed that this law was as fundamental as Newton's law of universal gravitation. Fechner was sure that his law would never be disproven, if only, as he said, because scientists would never agree on how to refute it. He was right only in part.

In 1961, Stanley Stevens published a paper entitled "To Honour Fechner and Repeal His Law." This paper was the result of Stevens' long-term research. He conducted his experiments differently than Fechner: Fechner's subjects had to distinguish very close stimuli, while Stevens' subjects evaluated the intensities of very different stimuli. Moreover, Fechner's subjects were not trained in advance; they evaluated stimuli by intuition, whereas Stevens' subjects were carefully taught how to evaluate stimuli correctly. Stevens' subjects' evaluations followed the power law:

$$\psi_s = cS^\beta, \qquad (3.2.2)$$

where S is the intensity of the physical stimulus, ψ_S is the subject's evaluation of the stimulus physical intensity, and β is a parameter depending on the type of stimulus. For example, for loudness, $\beta = 0.67$, for weight $\beta = 1.45$; c is the coefficient of proportionality.

We will demonstrate that both laws are embedded in the physical model, and that it clarifies the conditions under which these laws reveal themselves. The Weber-Fechner law appears in experiments where the subjects evaluate the intensities of their feelings in relation to stimuli, while the Stevens law holds where the subjects directly evaluate the physical intensity of a stimulus.

Let us begin our deduction of the laws of psychophysics. First of all, we will introduce theoretical analogues to the perceptive inputs to the physical model. These provide a connecting link between the physical world and the human mental domain. We will then show that the role of perceptive inputs can be played by the cylinders of heat engines. To deduce the laws of psychophysics, we will need three one-cylinder engines containing an ideal gas (Fig. 3.2.1).

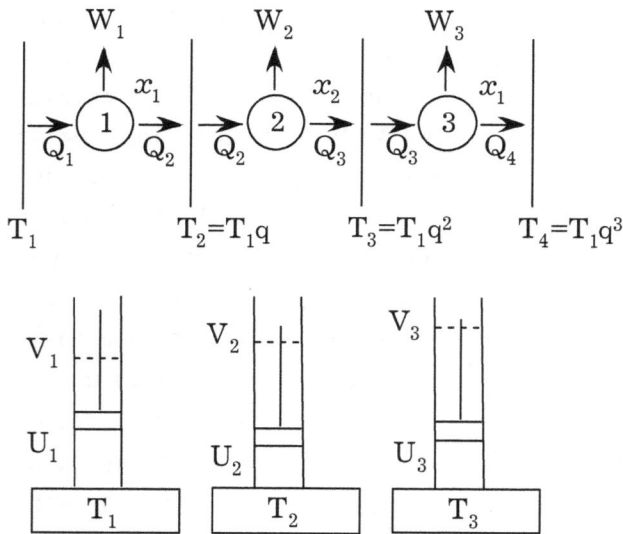

Fig. 3.2.1. Unit corresponding to the subject making
psychological evaluations (q is a positive number less than 1)

We will consider two subjects with descending and ascending reflexion, respectively. In descending reflexion, the first engine corresponds to the subject, the second engine to the subject's image of the self, and the third engine to the image of the self's image. The latter is the subject's *cognizant* image of the self: the subject knows that this is not he himself "in reality," but his image of the self. In ascending reflexion, the third engine corresponds to the subject, the second to the subject's image of the self, and the first to the subject's cognizant image. Work W_1, W_2, and W_3 performed by the engines are interpreted as the subjects' inner feelings.

For the subject with descending reflexion, W_1 is his immediate feelings, W_2 his feeling of his feelings, and W_3 his cognizant feeling. For the subject with ascending reflexion, the immediate feeling is W_3; W_2 is his feeling of that feeling, and W_1 is the subject's cognizant feeling.

We assume that the engines work cyclically. At the beginning of a cycle, the gas has the temperature of the heat source. The gas takes some heat from this source and expands isothermally. The details of how the engine returns the heat to the cold reservoir are not essential. The state of the ideal gas is described by the following equation:

$$PV = RT , \qquad (3.2.3)$$

where P is pressure, V volume, T temperature, and R the gas constant. Assume that $R = 1$. While expanding isothermally from V_a to V_b, the gas in a cylinder takes the following heat from the hot reservoir:

$$Q = \int_{V_a}^{V_b} \frac{T}{V} dV = T \ln \frac{V_b}{V_a} . \qquad (3.2.4)$$

The value

$$H = \ln \frac{V_b}{V_a} \qquad (3.2.5)$$

is the decrease of entropy as a result of receiving heat Q from the hot reservoir. The efficiencies of engines 1, 2, 3 are equal to x_1, x_2, x_1. We designate the initial volumes of gas in cylinders 1, 2, 3 as U_1, U_2, U_3, and the final ones as V_1, V_2, V_3, respectively. The work produced by the engines are

$$W_1 = x_1 T_1 \ln \frac{V_1}{U_1}, \tag{3.2.6}$$

$$W_2 = x_2 T_2 \ln \frac{V_2}{U_2}, \tag{3.2.7}$$

$$W_3 = x_1 T_3 \ln \frac{V_3}{U_3}. \tag{3.2.8}$$

The ratios of the initial and final volumes in the cylinders of engines 1 and 3 are related as

$$\left(\frac{V_1}{U_1}\right)^{T_1} = \left(\frac{V_3}{U_3}\right)^{T_2}, \tag{3.2.9}$$

where $T_1 > T_2$.

Now we begin our deduction of Stevens' law. In the descending reflexion, let the first engine cylinder be the perceptive input, and the third engine cylinder the cognizant output; $\frac{V_1}{U_1}$ is the intensity of the physical stimulus, and $\frac{V_3}{U_3}$ is the subject's cognizant evaluation of the intensity of the stimulus:

$$\frac{V_3}{U_3} = \left(\frac{V_1}{U_1}\right)^{\frac{T_1}{T_2}}. \tag{3.2.10}$$

In the ascending reflexion, the third engine cylinder is the perceptive input, and the first engine cylinder is the cognizant output. In this case, the cognizant evaluation of the intensity of the stimulus is

$$\frac{V_1}{U_1} = \left(\frac{V_3}{U_3}\right)^{\frac{T_2}{T_1}}.$$ (3.2.11)

The experimental subjects give evaluations proportional to $\frac{V_3}{U_3}$, in the descending reflexion, and to $\frac{V_1}{U_1}$ in the ascending reflexion.

Expressions (3.2.10) and (3.2.11) correspond to Stevens' law: in the first expression, the descending reflexion:

$$\frac{V_3}{U_3} = \psi_S, \quad \frac{V_1}{U_1} = S, \quad \frac{T_1}{T_2} = \beta;$$

in the second expression, the ascending reflexion:

$$\frac{V_1}{U_1} = \psi_S, \quad \frac{V_3}{U_3} = S, \quad \frac{T_2}{T_1} = \beta.$$

For the descending reflexion, $\beta = \frac{T_1}{T_2} > 1$, and for the ascending $\beta = \frac{T_2}{T_1} < 1$. The exact value $\beta = 1$ cannot be realized in this model.

Let us deduce Weber-Fechner's law. Consider the problem of how the subject's perceptual evaluations relate to his feelings. We connect feelings with the work produced by engines. Using (3.2.10) we can write (3.2.8) as

$$W_3 = x_1 T_2 \ln\frac{V_1}{U_1},$$ (3.2.12)

and (3.2.6) as

$$W_1 = x_1 T_2 \ln\frac{V_3}{U_3}.$$ (3.2.13)

Expression (3.2.12) corresponds to descending reflexion and (3.2.13) to ascending reflexion. The subjects give evaluations proportional to

W_3 and W_1. In this way we obtain logarithmic expressions relating the subject's feelings to the intensity of stimuli. These expressions correspond to the Weber-Fechner law (3.2.1). In (3.2.12)

$$W_3 = \psi_F, \quad V_1 = \varphi, \quad U_1 = \varphi_0,$$

and in (3.2.13)

$$W_1 = \psi_F, \quad V_3 = \varphi, \quad U_3 = \varphi_0.$$

The multiplier $x_1 T_2$ may be considered a constant.

Stevens' power law corresponds to direct evaluations of the intensity of physical stimuli. Weber-Fechner's logarithmic law corresponds to the subjects' evaluations of the intensity of their feelings about stimuli. One support to this interpretation is the fact that all Tempered musical scales are logarithmic.

3.3. Generating musical intervals

For millennia, the nature of musical intervals has puzzled musicologists. Physicist Richard Feynman wrote (1966, 50-1):

> We may ask ourselves if we better than Pythagoras understand now why only some tunes are pleasant to our ear. The general theory of aesthetics is not advanced more in our time than it was during Pythagoras.

Hermann Helmholtz (1885/1954) named the set of "attractive" intervals the Scale of Harmonical. They were not deduced formally but rather chosen based on their sounding (see McClain, 1976, 1987; Burns & Ward, 1982):

$$\frac{1}{1} \, \frac{15}{16} \, \frac{9}{10} \, \frac{8}{9} \, \frac{5}{6} \, \frac{4}{5} \, \frac{3}{4} \, \frac{2}{3} \, \frac{5}{8} \, \frac{3}{5} \, \frac{4}{7} \, \frac{5}{9} \, \frac{8}{15} \, \frac{1}{2} \tag{3.3.1}$$

The above ratios (except the octave, $\frac{1}{2}$, and unison, $\frac{1}{1}$) can be written as fractions of the type

$$\frac{n+1}{n+2} \quad \text{or} \quad \frac{1}{2}\frac{n+2}{n+1} \ , \tag{3.3.2}$$

and $\frac{2}{3}$ and $\frac{3}{4}$ may be written as either of the two.

Let us construct a model of the subject generating musical intervals using the physical model of psychological processes. We will do this with the help of a three-engine unit and suppose that, in the descending reflexion, the ratio of work for engines 1 and 2 is an integer, and, in the ascending reflexion, the ratio of work for engines 3 and 2 is an integer (see Fig. 3.3.1).

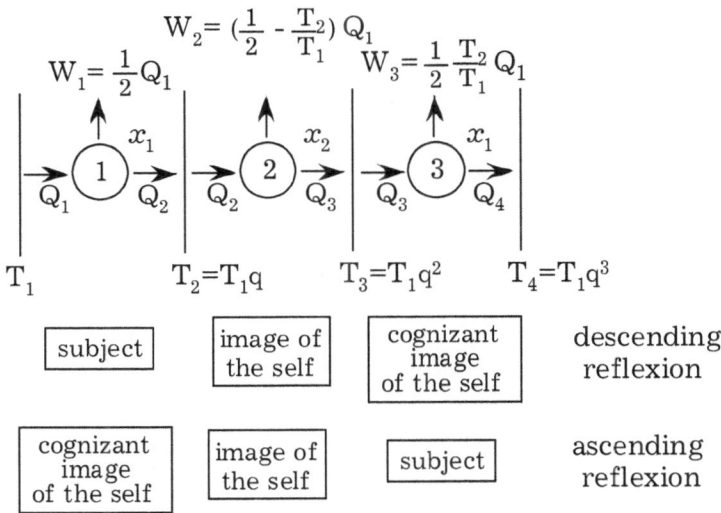

Fig. 3.3.1. Unit corresponding to a musician.
In the descending reflexion, the subject corresponds to the left engine,
in the ascending to the right one

With such integer relations, the subject and the subject's image of the self are in resonance (we say this metaphorically, because in classical thermodynamics there is no concept of time). The value of X_1 corresponds to a musical interval, and W_1, W_2, W_3 correspond to the subject's feelings (emotions). In the descending reflexion, W_1 is the subject's emotions, W_2 is the subject's image of the self's emotions, and W_3 is the subject's cognizant emotions. In the ascending reflexion, W_3 is the subject's emotions, W_2 is the subject's image of the self's emotions, and W_1 is the subject's cognizant emotions. We assume that the pressure toward the positive and negative poles is the same, i.e., $x_1 = \frac{1}{2}$.

Deduction of mathematical form of musical intervals

In the descending reflexion, engines 1 and 2 are in resonance, and the following expression corresponds to the musician:

$$\frac{W_2}{W_1} = \frac{\left(\dfrac{1}{2} - \dfrac{T_2}{T_1}\right)}{\dfrac{1}{2}} = \frac{1}{k+1} , \tag{3.3.3}$$

where k a positive integer. We derive from (3.3.3) that

$$\frac{T_1 - T_2}{T_1} = \frac{k+2}{2k+2} , \tag{3.3.4}$$

and

$$X_1 = \frac{x_1}{\dfrac{T_1 - T_2}{T_1}} = \frac{\dfrac{1}{2}}{\dfrac{k+2}{2k+2}} = \frac{k+1}{k+2} . \tag{3.3.5}$$

In the ascending reflexion, engines 2 and 3 are in resonance, and the following expression corresponds to the musician:

$$\frac{W_2}{W_3} = \frac{\left(\dfrac{1}{2} - \dfrac{T_2}{T_1}\right)}{\dfrac{1}{2}\dfrac{T_2}{T_1}} = R,$$ (3.3.6)

where R is either k, or $1/k$. We derive from (3.3.6) that

$$\frac{T_1 - T_2}{T_1} = \frac{R+1}{R+2}.$$ (3.3.7)

Hence,

$$X_1 = \frac{\dfrac{1}{2}}{\dfrac{R+1}{R+2}} = \frac{1}{2}\frac{R+2}{R+1}.$$ (3.3.8)

With $R=k$ the musician generates intervals of the type

$$X_1 = \frac{1}{2}\frac{k+2}{k+1},$$ (3.3.9)

and with $R=1/k$ he generates the intervals

$$X_1 = \frac{2k+1}{2k+2}.$$ (3.3.10)

It follows from (3.3.5), (3.3.9), and (3.3.10) that the musician generates intervals which can be written as

$$\frac{n+1}{n+2}, \quad \frac{1}{2}\frac{n+2}{n+1}.$$ (3.3.11)

We saw that the set of intervals separated by Helmholtz, except $\frac{1}{1}$ and $\frac{1}{2}$, consists of the intervals of the type (3.3.11). We will call intervals of this type *elite*.

Deduction of the set of musical intervals

There are countless fractions of the type (3.3.11). The set selected by Helmholtz consists of only 14 such intervals.

Interval d' is called an octave complement to interval d, if $dd' = \frac{1}{2}$. Most musical intervals in the set (3.3.1) have their octave complements there, except $\frac{8}{9}$ and $\frac{4}{7}$ whose octave complements are ($\frac{9}{16}$ and $\frac{7}{8}$). Let us add them to the Helmholtz set. In addition, we will also include consonance $\frac{6}{7}$ and its octave complement $\frac{7}{12}$ to (3.3.1) and obtain a new set which we name the *set of harmonical intervals*. All four added intervals are elite.

We will demonstrate now how these particular intervals are derived from the heat model. An interval is one of the simplest structural units in music. The next in complexity is a triad. It consists of three sounds which generate three intervals (Fig. 3.3.2).

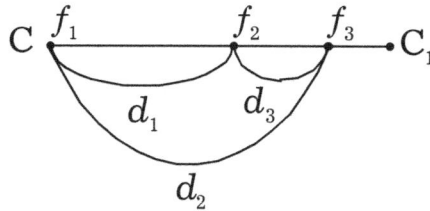

Fig. 3.3.2. Triad. Frequency f_1 corresponds to sound C;
frequencies f_2 and f_3 correspond to two other sounds;
d_1, d_2 and d_3 are the intervals between the sounds;

$$f_1 < f_2 < f_3 \leq 2f_1$$

The intervals are:

$$d_1 = \frac{f_1}{f_2}, \quad d_2 = \frac{f_1}{f_3}, \quad d_3 = \frac{f_2}{f_3}.$$

It is easy to see that

$$d_3 = \frac{d_2}{d_1}. \tag{3.3.12}$$

Let us consider a musician who can generate only elite intervals, i.e., intervals of the type (3.3.11). When d_1 and d_2 are elite, it may happen that d_3, calculated by (3.3.12), is not elite. Thus, to meet the condition that all three intervals be elite, only some sets of d_1, d_2, d_3 are solutions to equation (3.3.12). The elite intervals d_1 and d_2 will be called *compatible*, if the interval d_3 corresponding to them is also elite. The set of intervals compatible with an interval d^* will be called the interval d^* *suite*.

Statement 1. The set of harmonical intervals consists of unison ($\frac{1}{1}$), the intervals in the suite of the fifth ($\frac{2}{3}$), and their octave complements.

The fifth may be either interval d_2 or interval d_1.

If $d_2 = \frac{2}{3}$, then $d_3 = \dfrac{\frac{2}{3}}{d_1}$, which can be written as

$$\frac{k_3 + 1}{k_3 + 2} = \frac{\frac{2}{3}}{\frac{k_1 + 1}{k_1 + 2}}, \tag{3.3.13}$$

where k_1 and k_3 are positive integers.

If $d_1 = \frac{2}{3}$, then $d_3 = \dfrac{d_2}{\frac{2}{3}}$ or

$$\frac{k_3 + 1}{k_3 + 2} = \frac{\frac{1}{2} \frac{k_2 + 2}{k_2 + 1}}{\frac{2}{3}}. \tag{3.3.14}$$

Expressions (3.3.13) and (3.3.14) are Diophantine equations:

$$k_1 = \frac{k_3 + 5}{k_3 - 1}, \text{ where } k_3 \geq 2, \qquad (3.3.15)$$

and

$$k_2 = \frac{2k_3 + 8}{k_3 - 2}, \text{ where } k_3 \geq 3. \qquad (3.3.16)$$

Let us solve the Diophantine equation (3.3.15): if $m = k_3 - 1$, then

$$k_1 = 1 + \frac{6}{m}.$$

Since k_1 must be an integer, m may only be equal to 6, 3, 2, or 1. Whence,

$$k_1 = 2, 3, 4, 7.$$

Therefore, from all intervals of the type $\frac{k_1 + 1}{k_1 + 2}$, only the following four can be part of the suite of the fifth:

$$\frac{3}{4} \frac{4}{5} \frac{5}{6} \frac{8}{9} \qquad (3.3.17)$$

The Diophantine equation (3.3.16) can be presented as

$$k_2 = 2 + \frac{12}{m}, \text{ where } m = k_3 - 2.$$

Since k_2 is an integer, $m = 12, 6, 4, 3, 2, 1$. Hence,

$$k_2 = 3, 4, 5, 6, 8, 14$$

and the suite of the fifth includes the intervals of the type $\frac{1}{2} \frac{k_2 + 2}{k_2 + 1}$:

$$\frac{5}{8} \frac{3}{5} \frac{7}{12} \frac{4}{7} \frac{5}{9} \frac{8}{15}. \qquad (3.3.18)$$

By combining intervals (3.3.17) and (3.3.18) we obtain the suite of the fifth:

$$\frac{8}{9}\ \frac{5}{6}\ \frac{4}{5}\ \frac{3}{4}\ \frac{5}{8}\ \frac{3}{5}\ \frac{7}{12}\ \frac{4}{7}\ \frac{5}{9}\ \frac{8}{15}\ .$$

By adding to this set, unison ($\frac{1}{1}$) and the octave complements, which are not included in this obtained set, we obtain the final result:

$$\begin{array}{cccccccccccc}
\dfrac{1}{1} & \dfrac{8}{9} & \dfrac{5}{6} & \dfrac{4}{5} & \dfrac{3}{4} & \dfrac{5}{8} & \dfrac{3}{5} & \dfrac{7}{12} & \dfrac{4}{7} & \dfrac{5}{9} & \dfrac{8}{15} \\[2ex]
\dfrac{1}{2} & \dfrac{9}{16} & & \dfrac{2}{3} & & & \dfrac{6}{7} & \dfrac{7}{8} & \dfrac{9}{10} & \dfrac{15}{16} &
\end{array} \quad (3.3.19)$$

This set, as here derived, coincides with the set of attractive intervals.

So, we have proved that if the fifth ($\frac{2}{3}$) is used as a basic interval, a set of the corresponding intervals consists of all harmonical intervals. European music is based on the fifth. There are also countries where music is based on the fourth ($\frac{3}{4}$). For example, a developed musical system based on the fourth exists on the island of Java, it is called Pelog (Kunst, 1949). We can find a set of intervals based on the fourth in the same way as we found the set of intervals based on the fifth. It turned out that these sets coincide.

Statement 2. The set of harmonical intervals consists of the unison ($\frac{1}{1}$), the suite of the fourth ($\frac{3}{4}$) and their octave complements.

The following equations for the fourth correspond to (3.3.13) and (3.3.14) for the fifth:

$$\frac{k_3+1}{k_3+2} = \frac{\dfrac{3}{4}}{\dfrac{k_1+1}{k_1+2}}, \quad (3.3.20)$$

$$\frac{k_3 + 1}{k_3 + 2} = \frac{\dfrac{1}{2}\dfrac{k_2 + 2}{k_2 + 1}}{\dfrac{3}{4}}. \tag{3.3.21}$$

They can be rewritten as Diophantine equations:

$$k_1 = 2 + \frac{12}{m}, \tag{3.3.22}$$

this implies $m = 12, 6, 4, 3, 2, 1$;

$$k_2 = 1 + \frac{6}{m}, \tag{3.3.23}$$

and $m = 6, 3, 2, 1$.

We met similar equations earlier but here sets of k_1 and k_2 values are reversed in comparison with the case with the fifth. Equations (3.3.22) and (3.3.23) allows us to find the fourth's suite:

$$\frac{15}{16}\frac{9}{10}\frac{7}{8}\frac{6}{7}\frac{5}{6}\frac{4}{5}\frac{2}{3}\frac{5}{8}\frac{3}{5}\frac{9}{16}. \tag{3.3.24}$$

Let us add here unison ($\frac{1}{1}$):

$$\frac{1}{1}\frac{15}{16}\frac{9}{10}\frac{7}{8}\frac{6}{7}\frac{5}{6}\frac{4}{5}\frac{2}{3}\frac{5}{8}\frac{3}{5}\frac{9}{16}. \tag{3.3.25}$$

To the received set, we add the octave complements, which are not included in it:

$$\frac{1 \; 15 \; 9 \; 7 \; 6 \; 5 \; 4 \; 2 \; 5 \; 3 \; 9}{1 \; 16 \; 10 \; 8 \; 7 \; 6 \; 5 \; 3 \; 8 \; 5 \; 16}$$

$$\frac{1 \; 8 \; 5 \; 4 \; 7}{2 \; 15 \; 9 \; 7 \; 12} \quad \frac{3}{4} \quad \frac{8}{9}$$

(3.3.26)

Thus, we have obtained the set (3.3.26) of harmonical intervals based on the fourth equivalent to the set based on the fifth (3.3.19).

Statement 3. The set of harmonical intervals consists of the unison ($\frac{1}{1}$), the octave ($\frac{1}{2}$), the suite of the fifth and the suite of the fourth.

Our deduction of the set of harmonical intervals was based on the two laws of thermodynamics. This makes us think that this set is a fundamental attribute of the nature and we may use it for selection of the Universe models, as we use the anthropic principle.

Conclusion

In this book, we have tried to show that behind words like 'animacy', 'consciousness', 'mental experience', and 'subjectivity' there lies hidden a special "ideal element" which chooses the trajectory for the physical body at points of instability, following the fundamental laws of nature.

Here is a summary of our procedure. First, we constructed the mathematical model of a certain psychological process - bipolar choice - and tested the model. Then, we demonstrated that this model corresponds to a chain of ideal heat engines, whose functioning may be used to explain the phenomenon of bipolar choice. In addition, this chain of heat engines explains a number of other psychological phenomena unrelated to bipolar choice. The preceding suggests the hypothesis that the model of an ideal physical object, thus constructed, is linked to the process taking place in the *eidos*-navigator.

Such a hypothesis poses some additional questions. First, is the connection between the chain of ideal heat engines and some phenomena in the mental life of humans and animals only a coincidence? To answer this question, we have to analyze those features of formal theories that convince us their predictions are not coincidental. At the beginning of the twentieth century Niels Bohr constructed a model of the hydrogen atom. He represented it as a solar system consisting of a nucleus - the Sun - and an electron - a planet, which can leap from orbit to orbit and, in so doing, emit or absorb light quanta. This theory would not have been accepted by other physicists if it did not explain complicated patterns in the hydrogen spectrum. So, in saying that the prediction is not accidental, we are concerned not only with its correctness, but also with its complexity. Does our theory predict complex phenomena? Yes, it does. It predicts the set of harmonical intervals in music, which we obtained by solving Diophantine equations deduced from our thermodynamic model. This set consists of eighteen intervals, fourteen of which coincides with Helmholtz set of intervals.

Second, is the parallel between the mathematical model and the chain of heat engines appears due not to the existence of an *eidos*-navigator, but rather to some real physical process, described by these engines, that takes place in the physical brain of humans and animals? It is difficult to conceive of a heat process taking place in the brain with the significant and controlled changes in temperature required by the model. We may suppose, of course, that there is another physical process, not necessarily involving heat, but described by the same equations and with a different interpretation. Such a process is apparently unknown to science.

Third, does our model not contradict the law of conservation of energy? Before answering this question, let us look at the ontological schemes mostly used for an analysis of consciousness. They are of two types. In schemes of the first type, the physical world is closed causally and energetically. This means that consciousness, by itself, does not have any influence on the physical world. The best example is Epiphenomenalism, in which mental phenomena are viewed as the side effects of physical processes.

Schemes of the second type rely on dualistic frameworks, in which consciousness and the physical world are separate entities and consciousness may influence the physical world both causally and energetically. The physicist Squires (1996), in analyzing such dualistic models, cites Eccles:

> All attempts to formulate a dualistic hypothesis on brain-mind interaction are met with the strong criticism that such an hypothesis violates the conservation laws of physics. (Eccles, 1986, p.411)

The hypothesis described in this book is dualistic, of course. But it does not contradict the laws of conservation, because consciousness influences physical reality only under the condition that the body is in an unstable state at the moment when an infinitely small impact can change its trajectory. Note that the concept of 'infinitely small' has a meaning only in mathematics. Speaking of the physical world, we have to use 'operational unobservability', that is, to consider only impacts that do not reveal themselves in physical experiments.

One more question. Are there physical bodies that are not considered to be alive from the traditional point of view, but are animate according to our criteria? One example is Gaia. James Lovelock (1979) proposed the hypothesis that the Earth is a living organism which maintains stable conditions on the planet suitable for the existence of life. This organism influences non-biological components as well; for example, cloudiness is regulated to maintain necessary temperature for the biological mass. Such planetary processes contain a great number of bifurcation points and unstable zones; thus, in the framework of our hypothesis, the Earth may be considered an animate being.

Another example is the hypothesis that black holes are animate organisms (Lefebvre, Efremov, 2008). The mathematical model of a black hole shows that it has formal analogues for images of the self.

One more example. Attila Grandpierre (2004) proposed the hypothesis that the Sun is a living being.

The above examples relate to cosmic objects. We might also take an object of a much smaller scale, such as a soap bubble. Nonlinear processes take place in its membrane, and their dynamic trajectory contains indeterminate zones, so that it is very difficult to predict the moment when the bubble will burst. Similar behavior is demonstrated by linear and ball lightning.

A final example is the system of cracks in a solid body. This system contains points at which the extension of a crack is not determinate; thus, a system of extending tracks may be considered animate.

One may ask, does the *eidos*-navigator have an implicit goal that influences its choice of a dynamic trajectory? One possible answer: in generating a distribution of probabilities, the navigator aims to increase the number of ramification points and thus its ability to determine the body's behavior without violating physical laws of nature.

To find the nature of animacy is an urgent problem. Probably, it will be solved in the next few decades. The solution, I believe, will lie in the development of a new ontology elaborating the "materiality" of ideal objects.

References

Adams-Webber, J. & Benjafield, J. (1973).
    The Relations Between Lexical Marking and Rating Extremity in Interpersonal Judgment. *Canadian Journal of Behavioral Sciences*, **5**, 234-241.

Anderson, J. A. & Grice, J. N. (2009).
    Replicating a Simple Study of Asymmetry in Human Cognition. *Perceptual and Motor Skills*, **109**, 2, 577-580.

Barnett, J. D. (1915).
    The Operation of the Initiative, Referendum and Recall in Oregon. New York: The Macmillan Co.

Barrett, J. A. (1999).
    The Quantum Mechanics of Minds and Worlds. Oxford: Oxford University Press.

Baum, W. M., Schwendiman, J. W., and Bell, K. E. (1999).
    Choice, Contingency Discrimination, and Foraging Theory. *Journal of Experimental Analysis of Behavior*, **71**, 355-373.

Benjafield, J. & Adams-Webber, J. (1976).
    The Golden Section Hypothesis. *British Journal of Psychology*, **67**, 11-15.

Bohr, N. (1958).
    Atomic Physics and Human Knowledge. New York: John Wiley.

Bonanno, G. A. & Stillings, N. A. (1986).
    Preference, Familiarity, and Recognition After Repeated Brief Exposures to Random Geometric Shapes. *American Journal of Psychology*, **99**, 3, 403-415.

Burns, E. M., Ward, W. D. (1982).
    Intervals, Scales, and Tuning. In: Deutsch, D. (Ed.), *The Psychology of Music*, New York: Academic Press.

Butler, D. & Ranney, A. (1978).
    Referendums. Washington, D.C.: American Enterprise Institute for Public Policy Research.

Chalmers, D. (1996).
   The Conscious Mind. Oxford: Oxford University Press.
Eccles, J. C. (1979).
   The Human Mystery, London: Routledge & Kegan Paul.
Eccles, J. C. (1986).
   Do Mental Events Cause Neural Events Analogously to the Probability Fields of Quantum Mechanics? *Procedings of the Royal Society, Biological Sciences*, **227**, 411-428.
Eu, M. F. (1983a,b, 1985a,b, 1987, 1989a,b,c)
   Statement of Vote. Sacramento.
Fechner, M. F. (1860/1966).
   Elements of Psychophysics. New York: Holt, Rinehart and Winston.
Feynman, R. P., Leichton, R. B., & Sands, M. (1966).
   The Feynman Lectures on Physics. Reading: Addison-Wesley.
Grandpierre, A. (2004).
   Conceptual Steps Toward Exploring the Fundamental Nature of Our Sun, *Interdisciplinary Description of Complex Systems*, **2**, 1, 12-28.
Herrnstein, R.J. (1961).
   Relative and Absolute Strength of Response as a Function of Frequency of Reinforcement. *Journal of Experimental Analysis of Behavior*, **4**, 267-272.
Kelly, G. A. (1955).
   The Psychology of Personal Constructs. New-York: Harper & Row.
Kunst, J. (1949).
   Music in Java, Vol. 1 & 2, Maetinus Nijnoff.
Kunst-Wilson, W. R. & Zajonc, R. B. (1980).
   Affective Discrimination of Stimuli that Cannot Be Recognized. *Science*, **1**, 1-4.
Lefebvre, V. A. (1965).
   On Self-Reflexive and Self-Organizing Systems and their Study. In: *Systems and Structures Research*, p.61-68 (in Russian). English translation is given in Appendix II.
Lefebvre, V. A. (1985).
   The Golden Section and an Algebraic Model of Ethical Cognition. *Journal of Mathematical Psychology*, **29**, 289-310.
Lefebvre, V. A. (1987).
   The Fundamental Structure of Human Reflection. *Journal of Social*

and Biological Structures, **10**, 129-175.

Lefebvre, V. A. (1997).
   The Cosmic Subject. Moscow: The Institute of Psychology Press.

Lefebvre, V. A. (2004).
   Bipolarity, Choice and Entro-Field, *PROCEEDINGS: The 8-th World Multi Conference on Systemics, Cybernetics, and Informatics,* **IV**, 95-99.

Lefebvre, V. A. (2006a).
   Kosmichesky sub'ekt (the Cosmic Subject), third supplemented edition (in Russian). Moscow: Cogito-Center.

Lefebvre, V. A. (2006b).
   Research on Bipolarity and Reflexivity. Lewston, N.Y.: The Edwin Mellen Press.

Lefebvre, V. A., Efremov, Yu. N. (2008).
   Cosmic Intelligence and Black Holes. *World Future,* **64**, 563-576. First published in Russian: *Zemlia i Vselennaya,* 2000, No.5

Lefebvre, V. A. & Sanabria, F. (2008).
   Matching by Fixing and Sampling, *Behavioral Processes,* **78**, 204-209 (The paper is given in Appendix III).

Lefebvre, V. D. (1990).
   Choice without Criteria of Preference. In: Wheeler, H. (Ed.), *The Structure of Human Reflexion,* New York: Peter Lang.

Lovelock, J. E. (1979).
   A New Look at the Life on Earth. Oxford: Oxford University Press.

Mandler, G., Nakamura, Y., & VanZandt, B. J. S. (1991).
   Response Time. Oxford: Oxford University Press.

McClain, E. G. (1976).
   Pythagorean Plato. Stony Brook: Nicolas Hayes.

McClain, E. G. (1987).
   Comment on Vladimir Lefebvre's Tonal Automata. *Journal of Social and Biological Structures,* **10**, 204-212.

McGraw, K. M (1985).
   Subjective Probabilities and Moral Judgments, *Journal of Experimental and Social Psychology,* **14**, 501-518.

Myin, E. (2010).
   Unbounding the Mind, *Science,* **330**, 589-590.

Peitgen, H-O., Jurgens, H., Saupe, D. (1992).
 Chaos and Fractals. New Frontiers of Science. New York: Springer.
Penrose, R. (1989).
 The Emperor's New Mind. New York: Oxford University Press.
Poulton, E. S. & Simmonds, D. C. V. (1985).
 Subjective Zeros, Subjectively Equal Stimulus Spacing, and Contraction Biases in Very First Judgments of Lightness. *Perception & Psychopysics*, **37**, 371-404.
Ranney, A. (1978).
 The year of Referendum. *Public Opinion*, **1**, 5, 26-28.
Ranney, A. (1981).
 Referendums. 1980 Style. *Public Opinion*, **4**, 1, 40-44.
Ranney, A. (1983).
 The year of Referendum. *Public Opinion*, **5**, 6, 12-14.
Ranney, A. (1985).
 Referendums and Initiatives 1984. *Public Opinion*, **7**, 6, 15-17.
Ranney, A. (1987).
 Referendums and Initiatives 1986. *Public Opinion*, **9**, 5, 44-46.
Ranney, A. (1989).
 Election '88. Referendums. *Public Opinion*, **11**, 5, 15-17.
Regan, J. W. (1971).
 Guilt, Perceived Injustice, and Altruistic Behavior. *Journal of Personality and Social Psychology*, **18**, 1, 124-132.
Satinover, J. (2001).
 The Quantum Brain. New York: John Wiley & Sons, Inc.
Sayre, K. M. (1983).
 Plato's Late Ontology. Princeton: Princeton University Press.
Seamon, J. G., Brody, N., & Kauff, D. M. (1983).
 Affective Discrimination of Stimuli that Are not Recognized: Effects of Shadowing, Masking, and Cerebral Laterally. *Journal of Experimental Psychology: Learning, Memory, and Cognition*, **9**, 544-555.
Squires, E. (1990).
 Conscious Mind in the Physical World. Bristol and Philadelpha: Institute of Physics Publishing.
Stevens, S. S. (1961).
 To Honour Fechner and Repeal His Law, *Science*, **133**, 80-86.

Stevens, S. S. (1975).

Psychophysics. New York: John Wiley & Sons.

Stevens, S. S. & Galanter, E. H. (1957).

Ratio Scales and Category Scales for a Dozen Perceptual Continua. *Journal of Experimental Psychology*, **54**, 6, 377-411.

Appendix I
Abstract Heat Engines

In this Appendix we provide a concise description of classical thermodynamics as an abstract discipline.

Let us imagine that we can measure heat and work, that is, that we know how to associate these characteristics of physical processes with real numbers. Let heat be transformable into work and let work be transformable into heat, and let there exists a universal unit measure of both heat and work. At this point we introduce the concept of *energy* and consider work and heat as two possible forms of *energy passage*. For each pair of warm bodies, A and B, let us have a way to compare their heat and say either that "A is warmer than B" or that "B is warmer than A" or that "A and B are equally warm." We assume that this relation is transitive, i.e., if "A is warmer than B" and "B is warmer than C," then "A is warmer than C," and if "A and B are equally warm" and "B and C are equally warm," then "A and C are equally warm."

In this way, we have begun to construct thermodynamics without the concept of temperature. We assume, however, that there is a relation order on the set of all heated bodies and that for any two of them we can say which one is warmer or that they are equally warm. Now we introduce the second law of thermodynamics:

Work cannot be produced without the passage of heat
from a warmer body to a cooler one.

Since the second law of thermodynamics is formulated as a negative statement, and since we believe that everything not forbidden is permitted, then from this law follows the principle of receiving work from heat:

> *It is possible to construct a machine that would receive heat from a warmer body, give heat to a cooler body, and produce work W>0.*

The simplest heat machine that produces work is shown in Fig.1a. There are two heat reservoirs 1 and 2, where 1 is warmer than 2. In accordance with the principle of receiving work from heat, it is possible to construct a heat engine that would take heat Q_1 from the hot reservoir, give heat Q_2 to the cool reservoir, and produce work W. The *first law of thermodynamics*, or the *law of conservation of energy*, formulates relations between the quantities Q_1, Q_2 and W:

$$W = Q_1 - Q_2. \tag{1}$$

In accordance with the second law of thermodynamics, work W can be produced only if $Q_2 > 0$, this means that only part of the heat taken from the hot reservoir turns into work. This restriction does not exist for turning work into heat. The entire work produced by a heat engine can be turned into an equal amount of heat.

The engine is called *reversible* if it can work both of the ways shown in Fig.1: as 1a and as 1b.

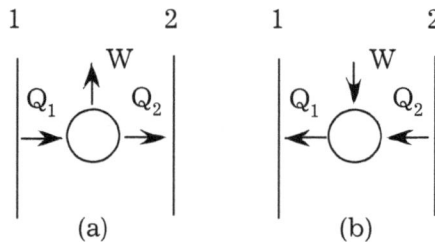

Fig. 1. Abstract heat engines. Vertical lines represent reservoirs of heat; reservoir 1 is hotter than 2. (a) Direct diagram: the engine takes heat Q_1 from reservoir 1, gives heat Q_2 to reservoir 2, and produces work W. (b) Reverse diagram: the engine receives work W from an external agent, takes heat Q_2 from the colder reservoir, and passes heat Q_1 to the hotter reservoir.

In other words, if the engine is reversible, we may take heat Q_2 from the cooler reservoir by spending work W and give it to the hotter reservoir, all in accord with the first law of thermodynamics, $Q_1 = W + Q_2$. The engine is called *non-reversible* if, after receiving work W from an external agent, we cannot take heat Q_2 from the colder reservoir and give heat Q_1 to the hotter one.   We assume that the heat reservoirs are so large that the engines' functioning does not change their "temperature." As we do not yet have a measure for temperature, we will say that transferring heat from a hot reservoir to a cold one does not change the order on the set of heat reservoirs. We will call such reservoirs *unchangeable*. The following variable

$$\rho = \frac{Q_1 - Q_2}{Q_1} \tag{2}$$

is called the *coefficient of efficiency*; it shows what portion of the heat coming from the hot reservoir is turned into work. The following statement holds:

Statement 1. For two unchangeable heat reservoirs, one of which is hotter than the other, there is no engine with a coefficient of efficiency higher than that in the reversible engine.

Proof. Consider Fig. 2. Engine 1 is reversible and engine 2 is some other engine (not necessary reversible). Both engines receive heat Q_1 from the hot reservoir. The reversible engine produces work W_1, and the second engine produces work W_2 (see Fig.2a). Suppose $W_2 > W_1$. The reversible engine corresponds to the reversed diagram, as we show it in Fig.2b. A portion of the work produced by engine 2 is spent to return heat Q_1 (taken by engine 2) to the hotter reservoir by passing work W_1 to engine 1. Since $W_2 > W_1$, the additional work $W_2 - W_1$ is produced without transferring heat from a hotter body to a cooler one: in the end, we returned the heat taken but produced work $W_2 - W_1 > 0$. This is forbidden by the second law of thermodynamics. Therefore, our supposition of $W_2 > W_1$ is wrong, and

$$W_1 \geq W_2 \qquad \blacksquare \qquad (3)$$

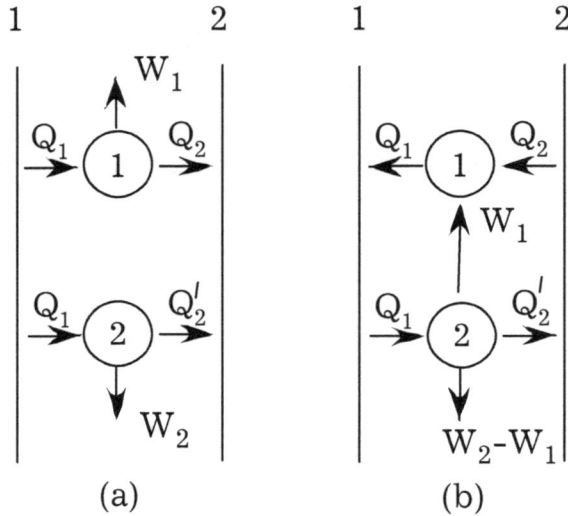

Fig. 2. Engine 1 is reversible; engine 2 may be either
reversible or non-reversible.

Carnot's theorem. All reversible engines working between the same heat reservoirs have the same coefficient of efficiency.
Proof. Let both engines in Fig.2 be reversible. It follows from statement 1 that since engine 1 is reversible, $W_1 \geq W_2$, and since engine 2 is reversible, $W_2 \geq W_1$. Thus, $W_1 = W_2$ and

$$\frac{W_1}{Q_1} = \frac{W_2}{Q_1} \qquad \blacksquare \qquad (4)$$

It follows from Carnot's theorem that, for a reversible engine, the coefficient of efficiency does not depend on the principle of its work nor on the fuel it consumes.

Let us now introduce the concept of *temperature*. Select one reservoir to be called *basic* and consider a set of other reservoirs, each of which is hotter than the basic one. Place a reversible engine between the basic reservoir and each of the others and take from them just as much heat as necessary for a reversible engine to release into the basic reservoir heat Q*. Heat Q* will be called the *unit of heat*. The quantity

$$T = \frac{Q}{Q^*},\qquad(5)$$

where Q is taken from a given reservoir, will be called its *temperature*. The temperature of the basic reservoir is equal to 1 (Fig.3).

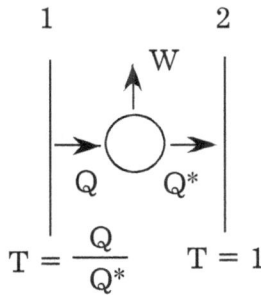

Fig. 3. Reservoir 2 is basic. Reservoir 1 is hotter than the basic reservoir. Q is the heat needing to be taken from reservoir 1 by a reversible engine to produce work W and release heat Q* to reservoir 2.

Statement 2. Reservoir 1 is hotter than reservoir 2, if and only if the temperature of reservoir 1 is higher than that of reservoir 2.
Proof.
(1) Necessity.
Let reservoir 1 be hotter than reservoir 2 and each engine, 1 and 2, release heat Q* into the basic reservoir (Fig. 4). Because engine 2 is reversible, diagram (a) in Fig. 4 can be transformed into diagram (b). Thus, the heat taken from the basic reservoir is equal to the heat returned to it. The work performed by the system in Fig. 4b is

Appendix I

$$W = W_1 - W_2 = (Q_1 - Q^*) - (Q_2 - Q^*) = Q_1 - Q_2. \tag{6}$$

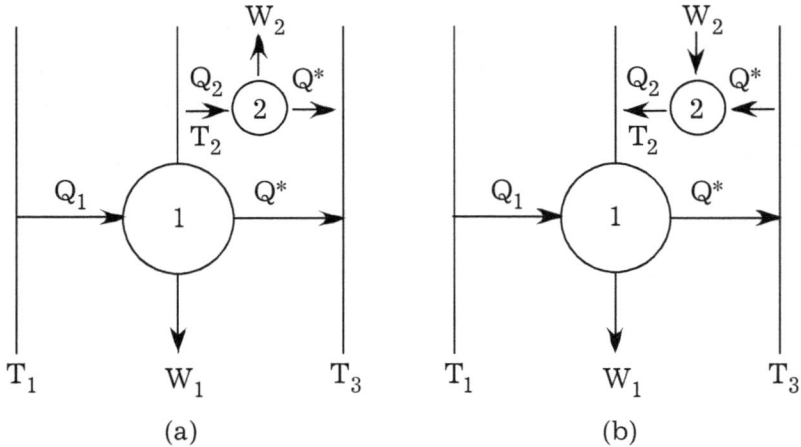

Fig. 4. Engines 1 and 2 are reversible. Reservoir 3 is basic.

Since the heat returned to the basic reservoir is exactly equal to the heat taken from it, the system can be considered as working between reservoirs 1 and 2, and since engines 1 and 2 are reversible, the whole system is also a reversible engine receiving heat Q_1 from reservoir 1 and returning heat Q_2 to reservoir 2. The coefficient of efficiency of a reversible engine is higher than zero, thus

$$W = Q_1 - Q_2 > 0, \tag{7}$$

hence

$$T_1 = \frac{Q_1}{Q^*} > \frac{Q_2}{Q^*} = T_2. \tag{8}$$

(2) Sufficiency. Let $T_1 > T_2$. Then (8) holds, therefore $W = Q_1 - Q_2 > 0$, which, according to the second law of thermodynamics, is possible only when reservoir 1 is hotter than reservoir 2.

∎

It also follows from (8) that, for the reversible engine, the coefficient of efficiency is

$$\frac{Q_1 - Q_2}{Q_1} = \frac{T_1 - T_2}{T_1}. \tag{9}$$

Since the coefficient of efficiency of the reversible engine is maximal under given conditions of functioning, then for any other engine working between reservoirs with temperature T_1 and T_2

$$\frac{Q_1 - Q_2}{Q_1} \leq \frac{T_1 - T_2}{T_1}. \tag{10}$$

Suppose another reservoir is chosen as basic. The temperature of reservoirs 1 and 2 on the new scale are $T_1{}'$ and $T_2{}'$ respectively. It follows from (9) that

$$\frac{T_1 - T_2}{T_1} = \frac{T_1' - T_2'}{T_1'} \tag{11}$$

or

$$\frac{T_2}{T_1} = \frac{T_2'}{T_1'}. \tag{12}$$

Thus, the ratio of temperatures for a pair of objects is the same on any scale.

Note that a formulation of the second law of thermodynamics may be given by expression (10) or by the equivalent expression

$$\frac{Q_2}{T_2} - \frac{Q_1}{T_1} \geq 0. \tag{13}$$

The statements formulated above, including the formulation of the second law of thermodynamics given in the beginning of this Appendix, can be deduced from (10). It follows that $Q_2 > 0$, i.e., producing work is impossible without transferring heat from a hotter body to a cooler one.

Consider now some correlations necessary for constructing the model. Engine M is placed between reservoirs with temperatures T_1 and T_2, where $T_1 > T_2$. It receives heat Q_1 from the hotter reservoir and releases heat Q_2 into the cooler one. The work produced by M is

$$W_1 = Q_1 - Q_2. \tag{14}$$

In addition, a reversible engine is placed between the same reservoirs; it produces work

$$W_0 = Q_1 \frac{T_1 - T_2}{T_1}. \tag{15}$$

Theoretically, this work is the maximum possible for the given ratio of temperatures in the reservoirs. The quantity

$$\Delta W_1 = W_0 - W_1 \tag{16}$$

is called the *lost available work*. If engine M is non-reversible,

$$\Delta W_1 > 0. \tag{17}$$

The lost available work is the energy lost by a heat engine due to its imperfection. It is the additional work which M could produce if it were reversible. It follows from (16) that

$$\Delta W_1 = Q_1 \frac{T_1 - T_2}{T_1} - (Q_1 - Q_2) = Q_2 - \frac{T_2}{T_1} Q_1 = T_2 \left(\frac{Q_2}{T_2} - \frac{Q_1}{T_1} \right). \tag{18}$$

The quantity

$$\Delta H = \frac{Q_2}{T_2} - \frac{Q_1}{T_1} \tag{19}$$

is called the system's *entropy change* as a result of producing work W_1. Therefore,

$$\Delta W_1 = T_2 \Delta H. \tag{20}$$

Appendix II
On self-reflexive and self-organizing systems[1]

At the outset let us highlight and distinguish two processes:
a) the representation of an object as a system and
b) the construction of an object according to a plan.

1. Representation of an object as a system. In order to solve certain cognitive problems we have to picture an object as divided into its elements. The picture must indicate the ties and relations which turn this "segmentation" into "wholeness". The elements, ties, and relations may be chosen differently according to the 'standards' which a researcher uses to reflect the object and which determine its structure.

Let us take a set of dots that have to be counted.

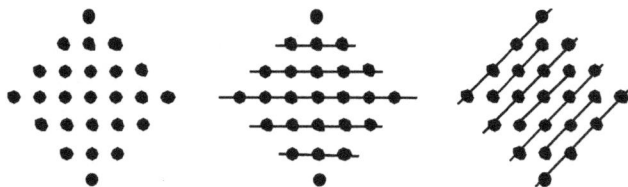

Fig. 1

Depending on the mathematical tools available to the researcher, the object can be represented as a system of either horizontal or inclined lines. It is obvious that these representations of the system are different, and yet it makes no sense to ask which one represents the object more precisely.

---

[1]SYSTEMS AND STRUCTURES RESEARCH PROBLEMS,
Conference Proceedings, Moscow, 1965 (in Russian)

2. Constructing an object as a system. In solving practical problems, special objects are constructed from a set of parts according to a particular "plan-image" which guides a "designer" in transferring the "image" to another sphere. He realizes the plan in a different medium. The object obtained possesses a structure by virtue of its being created.

3. Analysis of the constructed object. The researcher may be faced with a special problem: determining an object's inherent structure. The object, however, as an object of study, takes on the structure of its image (because "to see the structure of an object means to represent that object"). Therefore, it is necessary to choose a system representation for the object that reflects precisely the structure obtained by it. This suggests a natural principle for the researcher: to borrow the "plan" that guided the designer and use it in a new function - as a means for representing the system.

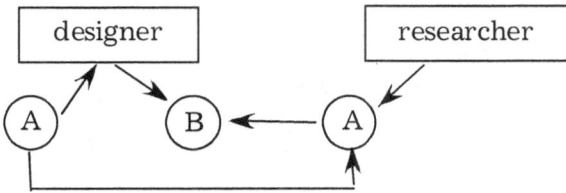

Fig.2

Let us take an ancient text, which initially looks like a random conglomeration of lines and dots:

Fig. 3

The task is to decipher it. A linguist would divide the text into distinct elements based on the objective writing standards of an ancient people. The linguist must reconstruct the standards and, from that point of view, represent the conglomeration of lines and dots as a system. A wrong initial representation would condemn all subsequent work to failure, and only one representation will allow the task to be solved correctly.

4. Distinguishing between organization and systemicity. Must a "designer" be a Man (or something created by man)? We have no reason to insist that this be so. Any mechanism that carries out a "plan" by giving structure to material can be regarded as a "designer." This leads us to concepts of organization and of an organizing system that is comparable to a human researcher in its "perfection". A system consisting of two elements A and B and a special determining mechanism which structures element B according to project A will be called an *organizing system*.

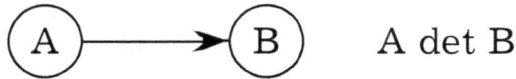

A det B

Fig. 4

The structure acquired by element B in this system will be called its *organization*, as opposed to the structure given the element by a human researcher, which will be called its *systemicity*. Such a scheme enables us to introduce the concept of "organization" as characterizing an element of a system but not the whole. To define this concept for the latter we must introduce the concept of a *self-organizing system*.

5. Self-organizing system. A system in which one element functions as the plan for the whole and a particular mechanism structures the whole according to the "plan" will be called a *self-organizing system*. The organization of the whole is that structure which is generated by the realization of the plan.

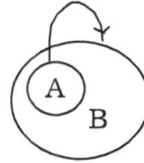

A *det* B
A ∈ B

Fig. 5

6. Principle of borrowing. A self-organizing system can be studied just like any other object. In this case, the choice of how the system should be represented is completely up to the researcher. But if we seek to study a self-organizing system *qua* self-organizing, the choice is determined by the system itself. The researcher must extract the system's design from the system, include it (usually with some modification) in the set of his *means for system representation*, and examine the system as if from the point of view of the system itself. The principle of representing a system using means extracted from the system itself will be called the *principle of borrowing*.

7. Degree of organization. The deviation of the system's structure from its plan (system dissonance) can be taken as a measure of the system's organization. The greater the deviation, the less the organization. This allows us to introduce a concept of *degree of organization* without relying on an absolute universal measure, whose function is often assigned to entropy.

8. Systems - configuroids. Consider the process of constructing an electronic device. Omitting insignificant details, we can describe this process in Figure 6. Two different structures are materialized in an electronic device; it has two organizations, both embedded in it. Systems which involve such "symbiosis" of different structures will be called *configuroids*. A special class is constituted by configuroids with antagonistic structures. To represent a configuroid, it is necessary to construct a system of system representations, that is, a *configurator*.

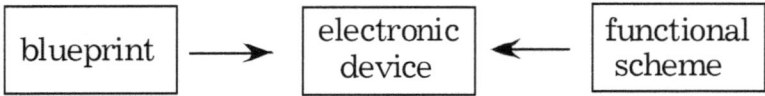

blueprint	$\longrightarrow$	electronic device	$\longleftarrow$	functional scheme

Fig. 6

9. Self-reflexive system. A system, one element of which functions as a representation of the whole, will be called a *reflexive* system. A system which is both reflexive and self-organizing will be called *self-reflexive*. Reflecting the whole in one of its elements will be called *reflexion*.

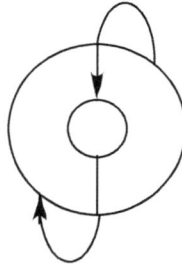

Fig. 7

10. A group as a self-reflexive system. The concept of a self-reflexive system makes it possible to explain some aspects of the functioning and evolution of groups of primitive people. In particular, it makes it possible to construct a mechanism for the origin of individual reflexion. Let us construct a model of a primitive group.

We distinguish between the "leader" and an "ordinary member" of the group. The leader acts as a "designer" of group situational structures. This is his one and only function. Every ordinary member possesses a number of working procedures not internally connected with one another; they can be connected into a sequence only by the leader's agency.

All the members of the group deal with reality, but the leader deals with a special kind of reality - that of the group itself. The leader is apart from the group and stands above it. He can fulfill the function of a designer if he assimilates this reality by mapping it onto a special tablet, transforms that representation into a plan, and then executes this plan.

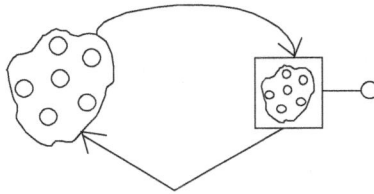

Fig. 8

The tablet must represent ordinary members of the group, the object assimilated by them, and special notations of the procedures performed by these members.

Thus, there are two types of activity:
1) work operations performed by ordinary members, and
2) special operations related to a particular object, a group, and performed by a leader using a special semiotic means, that is, a tablet.

It would seem that a "herd" turns into a group at the moment when it becomes self-reflexive, that is, when semiotic means appear for planning the activity of a group as a whole.

11. Small groups. Let us limit the number of ordinary members in a group and leave the number of necessary working procedures unchanged. In a small group the leader must perform some ordinary functions in addition to his functions as leader. Thus, while solving certain problems the leader must map himself onto the tablet as a special material substitute for the self, together with the other ordinary group members (for example, in allocation problems). From this point a new phenomenon arises. Both ordinary operations and specific organizational activity are conflated in the person of the leader.

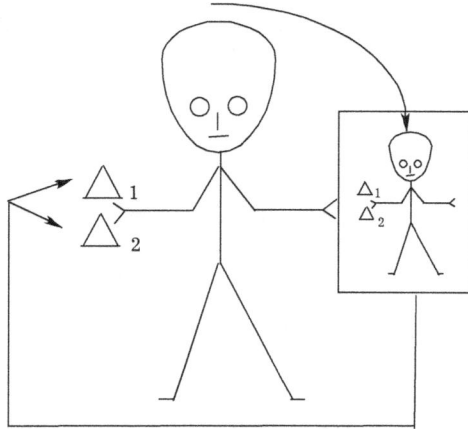

Fig. 9

For the first time, individual activity becomes *organized*. Now the mechanism that previously functioned within the framework of a group is transferred to individual activity. The leader turns into a *self-reflexive system*.

In the beginning, the system for controlling the group (signals) has to be a component of the leader's individual activity. But since there are no spatial distances needing to be overcome in this system, the signals in individual activity become irrelevant, and a direct connection between the tablet and the leader's working procedures is established.

Apparently, the connection of two different types of activity through the creature's mapping itself onto a tablet is the moment when individual reflexion appears. The "self" appears as an external material substitute for the leader. Initially, the loss of the material substitute is a loss of reflexion (self-consciousness). Only later, when the object-substitute comes to be reflected physiologically, does it evolve into a "head."

Individual consciousness cannot appear in the "head." To explain its appearance, one must study the structure of group activity and the evolution of symbolic systems. The problem of the origin of man, as well as that of the origin of primitive society, are essentially semiotic problems.

Appendix III
Matching by Fixing and Sampling: A Local Model Based on Internality[1]

Vladimir A. Lefebvre
Federico Sanabria

ABSTRACT

Undermatching and overmatching in concurrent schedules of reinforcement have been traditionally described as changes in the slope of the Generalized Matching Law function. More recently, Baum, Schwendiman, and Bell (1999) suggested that deviations from strict matching may be better described as following a policy of mostly *fixing* on the preferred schedule, and occasionally *sampling* the alternative schedule. So far, no model of local performance predicts the global outcome of this policy. We describe one such model based on parsimonious assumptions of the internal state of the organism and mechanisms of reinforcement attribution. Formally, the model is analogous to the Axiom of Repeated Choice (Lefebvre, 2004).

Keywords: Concurrent schedules, choice, generalized matching law, internality, visit patterns, fix and sample, Axiom of Repeated Choice, contingency-discriminability model

---

[1]Behavioral Processes, **78**, pp.204-209

The Matching Law is a fundamental concept in behavioral choice theory (Herrnstein, 1961; for a review, see Davison and McCarthy, 1988). It states that the allocation of behavior across alternatives matches the distribution of obtained reinforcers (De Villiers and Herrnstein, 1976); mathematically, it may be expressed as

$$\frac{B_1}{B_1 + B_2} = \frac{r_1}{r_1 + r_2} ,$$
(1)

or in its equivalent ratio form (Baum and Rachlin, 1969)

$$\frac{B_1}{B_2} = \frac{r_1}{r_2} .$$
(1')

where B_1 and B_2 are the rates of responding on each alternative, and r_1 and r_2 are the rates of reinforcement obtained from the corresponding alternatives.

Extensive empirical research on animal and human choice has shown that experimental subjects systematically deviate from "strict" matching (Davison and McCarthy, 1988). To account for these deviations, Baum (1974) suggested a generalization of Equation 1' known as the Generalized Matching Law (GML):

$$\log\frac{B_1}{B_2} = a \log\frac{r_1}{r_2} + \log b$$
(2)

Parameter a represents the *sensitivity* of choice to relative rate of reinforcement (Lobb and Davison, 1975), and b is the *bias* towards schedule 1 that is not due to r_1 or r_2. If $a > 1$, rates of reinforcement are *overmatched* by choice; if $a < 1$, rates of reinforcement are *undermatched*. Undermatching is prevalent across studies (Davison and McCarthy, 1988), although overmatching is occasionally observed (Aparicio, 2001).

Much work has been done in the last 3 decades to specify local processes that would yield matching in the long run (Davison and Jenkins 1985; Herrnstein, 1982; MacDonall, 1999; Staddon and

Motheral, 1978; Wearden, 1983). Consistent with the global pattern described by GML (Equation 2), these local models assume that the function that relates choice to reinforcement—the *matching function*—is continuous. Recent evidence, however, has questioned this assumption.

Discontinuity in the Matching Function: Fix-and-Sample Patterns of Choice

The Matching Law has been typically demonstrated using concurrent variable-interval variable-interval (Conc VI VI) schedules of reinforcement. In this procedure, subjects continuously choose between two sources, each of which delivers rewards at a programmed rate but with no periodicity. By examining local patterns of choice in Conc VI VI, Baum et al. (1999) uncovered a "fix-and-sample" pattern in pigeon's choices: Pigeons generally responded on the alternative that yielded more reinforcers (*rich* schedule) and occasionally made a few responses on the alternative that yielded fewer reinforcers (*lean* schedule); whereas responding on the rich schedule was sensitive to relative rate of reinforcement, responding on the lean schedule was not. This fix-and-sample pattern has also been detected in rhesus monkeys (Lau and Glimcher, 2005) and rats (Aparicio and Baum, 2006).

The fix-and-sample pattern implies a discontinuity in the matching function where schedules change from lean to rich and vice versa. To illustrate this point, consider what happens when a schedule switches from being lean to being rich. While the schedule was lean, the duration of each visit was fixed, say, to 4 keypecks on the average, regardless of its relative rate of reinforcement. Now that it is the rich schedule, the duration of each visit is sensitive to rate of reinforcement. When the rich schedule is only slightly richer than the lean schedule, sensitivity to rate of reinforcement may drop visit lengths below 4 keypecks or may boost them above that number, yielding respectively a step down or up in the matching function.

To account for the fix-and-sample pattern observed in their experiment and for the discontinuity that it implies for the matching function, Baum et al. (1999) suggested that GML operates with no sensitivity parameter on the ratio of reinforcers obtained from the rich (r_R) and lean (r_L) schedules (see also Baum, 2002):

$$\frac{B_R}{B_L} = b_R \frac{r_R}{r_L} \tag{3}$$

or, equivalently,

$$\log \frac{B_R}{B_L} = \log \frac{r_R}{r_L} + \log b_R \tag{3'}$$

where B_R and B_L are the rates of responding on the rich and lean schedules, respectively, and b_R is the bias towards the rich alternative. Note that the value of b_R can be greater than 1, smaller than 1, or equal to 1, so $\log b_R$ can be either positive or negative or zero. In this formulation the matching function intercepts the ordinate at $\log b_R$. At this point, $r_R = r_L$, so the distinction between rich and lean is meaningless; if r_R is further reduced, however, the rich schedule to the right of the ordinate becomes the lean schedule, and vice versa. Thus, when Equation 3' is plotted with the ratio of reinforcement rates ($\log[r_1/r_2]$) on the abscissa, a discontinuity is observed at the ordinate. Two possible discontinuities—step down or up—are described by the ideal matching functions in the top panels of Figure 1; the step-down pattern (left) corresponds to undermatching, whereas the step-up pattern corresponds to overmatching.

To the left of the ordinate in the top panels of Figure 1, alternative 1 is lean and alternative 2 is rich; to the right of the ordinate, alternative 1 is rich and alternative 2 is lean. Tracing Equation 3' from left to right, the function breaks at $b'_{R=2}$, which is b_R when alternative 2 is rich. The function then continues from $b'_{R=1}$, which is b_R when alternative 1 is rich. From equation 3, it may be shown that $b'_{R=1}$

$= -b'_{R=2}$, as illustrated by the symmetry of the discontinuity around the origin (Figure 1, top panels). A more general form of the model, which we call the Fix-and-Sample (FS) model, assumes a bias towards alternative 1 (b_1), such that

$$b_{R=1} = b_1 b'_{R=1}, \quad b_{R=2} = b_1 b'_{R=2} . \tag{4}$$

Coefficient b_1, allows the matching function to be shifted up or down—not just symmetrically around the x-axis—to fit data. The bottom panels of Figure 1 illustrate the fit of the model to data, based on the performance of one rat with two levels of effort for changing over alternatives (Aparicio, 2001). Note that the discontinuities in the fitted functions are not vertically centered on zero, but on a negative number, indicating that b_1 was negative—choice was biased toward alternative 2.

It is not obvious how a discontinuity in behavior between richer and leaner schedules can be incorporated to continuous-function local models of matching without violating their fundamental assumptions. The FS model (Equations 3' and 4) describe global patterns of behavior allocation, but it does not explain why these equations must hold and does not specify local choice mechanisms whose aggregated operation would yield fix-and-sample patterns. We propose a hypothetical local choice mechanism that is consistent with the FS model. The mechanism is based on parsimonious assumptions of the internal state of an organism, which are specified by the formal relation between the system's internal state, the environment's influence, and the probability with which the system chooses each alternative.

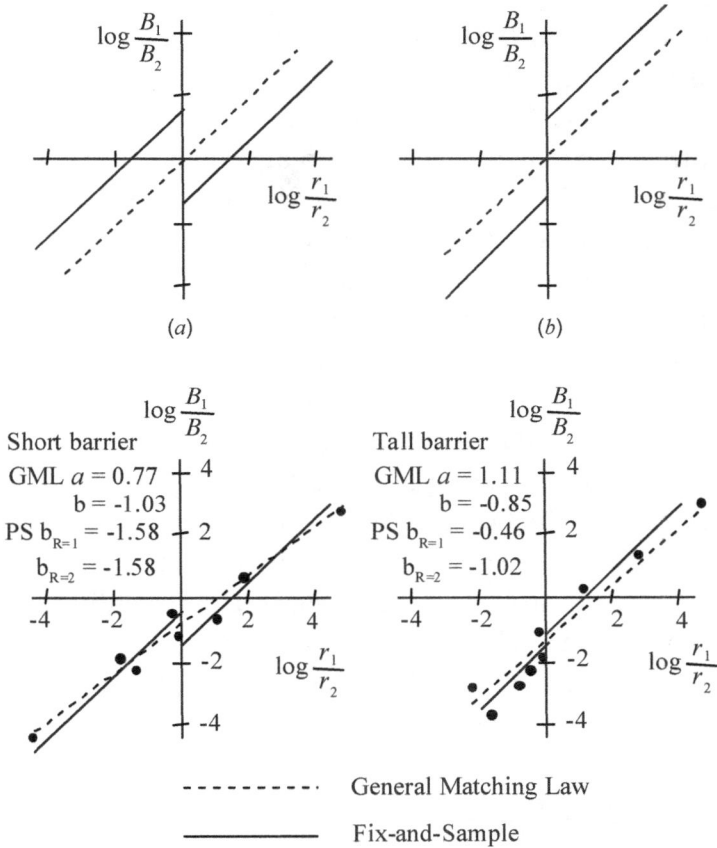

Figure 1. *Top panels.* Ideal patterns given by Eq. 3'. The pattern on the left panel corresponds to the case when $\log b_R < 0$, i.e., undermatching. The pattern on the right panel corresponds to the case when $\log b_R > 0$, i.e., overmatching. The dotted line corresponds to $\log b_R = 0$. *Bottom panels.* Illustrative fitting of the Generalized Matching Law (GML; Equation 2) and the Fix-and-Sample model (FS; Eqs. 3' and 4) to performance in concurrent variable-interval variable-interval (Conc VI VI) schedules of reinforcement. Relative rates of responding are shown as a function of relative rates of reinforcement in one rat (#62) when response levers were separated by a 30.5 cm barrier (left panel) and by a 45.7 cm barrier (adapted from Aparicio, 2001). The dashed line is the best fitting form of GML and the solid lines are the best fitting form of FS. Fitted parameters are displayed for each model.

A Model of Bipolar Choice

This model was first introduced to explain some phenomena in human moral choice (Lefebvre, 1982, 1992). Consider the following situation. A person is facing an alternative: to tell the truth or to lie. Let the truth be "good" for this person and to lie "bad." In addition, for telling the truth the person would receive $10 and for lying $10,000. This is an example of a situation in which a choice has two aspects, one moral and one utilitarian. In the moral aspect, the alternatives are bipolar. One of them can be called the positive pole and the other the negative pole. In the utilitarian aspect, the alternatives are assigned with numbers that correspond to their utility. In this particular example, the negative pole (lie) is more profitable than the positive pole (truth). We cannot be certain of which alternative will be chosen, because polarity and utility are inconsistent with each other. This is a situation akin to the complex ambivalence self-control scenario described by Rachlin (2000).

The formal model of bipolar choice was initially constructed for predicting human choice in situations of this type; it may be represented in the following equation:

$$X = \frac{x_1}{x_1 + (1 - x_1)x_2}, \quad 0 < x_1 \leq 1, \quad 0 \leq x_2 \leq 1 \ . \tag{5}$$

In this equation, X is the probability of choosing the positive pole, x_1 is the relative utility of the positive pole, and x_2 is a parameter characterizing the subject's inner state, the value of which is determined by a larger context. If $x_2=0$ then $X=1$, that is, choice is completely based on the polarity of the alternatives; if $x_2=1$ then $X = x_1$, that is, choice is completely based on the utility of the alternatives. Thus, x_2 is the relative decision weight given to local and global considerations.

After the model of bipolar choice had been constructed, it was found that it could also make predictions beyond the area of moral choice. For example, it could explain a few psychophysical phenomena, among them the non-linear relation between magnitude and categorical estimations of the same physical stimuli (length, weight, duration, area) (Lefebvre, 1992). The model shed a new light on the asymmetry in evaluations given by people to their acquaintances in using constructs of the type strong-weak, fast-slow, etc. Experimental data have demonstrated that the frequency of choosing a positive adjective was equal to 0.62, but not 0.5 as was expected; the model explained this shift (Adams-Webber, 1997; Lefebvre, 1980). These results led to the notion that the model expressed by Equation 5 may describe not only human choice but animal choice as well.

We hypothesized that choices in Conc VI VI have two aspects, which may be called *utilitarian* and *positive-negative*. The utilitarian aspect relates to the immediate preferences of the animal at a local scale, and positive-negative aspects to the animal behavior in a larger, global, time scale. We can rewrite Equation 5 as

$$\frac{1-X}{X} = x_2 \frac{1-x_1}{x_1} \ , \tag{6}$$

which allows us to see a parallel between the model of bipolar choice and Equation 3. Lean and rich alternatives may be assigned with positive and negative polarity—the assignment criteria will be discussed further below. For illustration, let the lean alternative be the positive pole and the rich alternative the negative one:

$$X = \frac{B_L}{B_R + B_L}, \quad x_1 = \frac{r_L}{r_R + r_L} \ . \tag{7}$$

After substituting these values in Equation 6 we obtain

$$\frac{B_R}{B_L} = x_2 \frac{r_R}{r_L} \ . \tag{8}$$

Equation 8 is analogous to Equation 3.

A New Version of the Model of Bipolar Choice

In this section, we demonstrate that Equation 8 may be derived from a few assumptions regarding local choice. Let us first represent the behavior of an organism in a situation of bipolar choice as the following function:

$$X = \Phi(x_1, S), \quad 0 < X \le 1, \quad 0 < x_1 \le 1, \quad S \ge 0 \cdot \qquad (9)$$

The values of variables X and x_1 are interpreted as in the old model: X is the long-term probability of an organism choosing the positive alternative, and x_1 is the instantaneous probability with which the local environment instigates the organism to choose the positive alternative. S is an internal variable that expresses the organism preference for the positive alternative at a global scale. Let us assume that the local impact of reinforcement is already established (e.g., animals are predisposed to prefer more immediate rewards), whereas global preference must be acquired. We assume that with x_1 being constant and S growing, the probability of choosing the positive pole, X, grows. If there is no long-term preference, $S = 0$, the probability of choosing a pole is completely determined by local influence:

$$\Phi(x_1, 0) = x_1 \cdot \qquad (10)$$

The probability with which the system chooses the positive alternative when the internal variable is equal to S will be designated as X_S. To find function $\Phi(x_1, S)$, we invoke one assumption, which we called the Axiom of Repeated Choice:

> When the internal variable grows from S to $S + \Delta S$ ($0 < \Delta S < 1$; ΔS is considered small) and x_1 does not change, the procedure of choice is as follows. First, the system makes a *preliminary* choice with the probability of choosing the positive alternative equal to X_S. If the positive alternative is chosen, the system realizes its choice. If the negative alternative is chosen, then, with a small probability equal

to ΔS, the system *cancels* its choice and *repeats* the procedure of choice (with the probability of choosing the positive alternative equal to X_S). The result of the repeated choice is realized no matter which alternative is chosen. (Lefebvre, 2004).

The empirical basis of the Axiom of Repeated Choice is incomplete, and thus it should be deemed as a hypothesis. Nonetheless, the phenomenon of repeated choice was observed in subjective estimation of stimulus intensity on a linear scale (Poulton et al., 1968). Subjects were presented with a stimulus and had to rate its intensity on a 1-100 scale. The experimenters found that after marking the scale, subjects often crossed it out and repeated the procedure of choice anew. Lefebvre (2006) demonstrated that these results could be analyzed in a scheme of choice between positive and negative alternatives.

The Axiom of Repeated Choice is depicted in Figure 2.

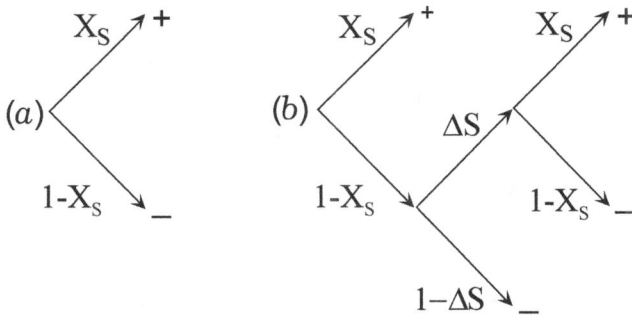

Figure 2. Decision trees derived from the Axiom of Repeated Choice (a) when the value of the internal variable remains constant at S, and (b) when the value of the internal variable changes to $S + \Delta S$.

When there is no change in the internal variable S, choices are made according to decision tree (a): the positive alternative is chosen with probability X_S and the negative alternative with probability $1 - X_S$. When S grows by ΔS, decision tree (b) is applied. The initial choice in tree (b) is similar to the choice in tree (a), but if the negative alternative is chosen, there is a small probability, ΔS, that the choice will be reconsidered.

It follows from tree (b) in Figure 2 that the probability of choosing the positive alternative when the internal variable changes $(X_{S+\Delta S})$ is the sum of the probability of choosing the positive alternative in the preliminary stage (X_S) and of the joint probability of choosing the negative alternative in the preliminary stage $(1 - X_S)$, reconsidering the choice (ΔS), and finally choosing the positive alternative (X_S):

$$X_{S+\Delta S} = X_S + (1 - X_S)\Delta S X_S \ . \tag{11}$$

As preference for the positive alternative at a global scale is acquired, the internal variable stabilizes and $\Delta S \rightarrow 0$. By considering X_S as a differentiable function with argument S and assuming $\Delta S \rightarrow 0$, we obtain the following differential Equation:

$$\frac{dX(S)}{dS} = (1 - X(S))X(S) \ . \tag{12}$$

After solving Equation 12 under condition $X(0) = x_1$, we find that

$$X = \Phi_{x_1}(S) = \frac{x_1}{x_1 + (1 - x_1)e^{-S}} \ . \tag{13}$$

Equation 13 corresponds to Equation 5 for bipolar choices, if $x_2 = e^{-S}$. Analogously to Equations 5 and 8, Equation 13 can be transformed to

$$\frac{B^-}{B^+} = e^{-S}\frac{r^-}{r^+} \ , \tag{14}$$

where '+' corresponds to the positive alternative and '-' to the negative one. By finding the logarithms of the right and left sides of Equation 14, we arrive to the final Equation:

$$\ln\frac{B^-}{B^+} = \ln\frac{r^-}{r^+} - S \ . \tag{15}$$

This equation describes the relation between the frequencies of choosing positive and negative alternatives, the frequencies of their reinforcements, and the internal variable.

Let us now compare the Equations 3' (from the FS model) and 15. If it is assumed that the logarithm base in Equation 3' is e and $\ln b_R = -S$, and considering that $S \geq 0$, it follows that the richer and leaner schedules are negative and positive, respectively. This is the scenario where matching is biased toward the leaner schedule and thus choices undermatch reinforcement. The polarity of the alternatives may be different, in which case $\ln b_R = S$, which implies that $\ln b_R \geq 0$. This is the scenario where matching is biased toward the richer schedule and thus choices overmatch reinforcement.

The Meaning of Polarity

What does it mean for a lean or rich schedule to be "positive" or "negative"? In the original model of bipolar choice, polarity indicated the desirability of an alternative in a larger, ethical context. Prior research has shown that pigeons may become sensitive to global contingencies of reinforcement that are inconsistent with local contingencies, as when an alternative is preferred in the long run but not immediately (Heyman and Tanz, 1995; Sanabria et al., 2003). If polarity of alternatives signaled long term contingencies, it would be expected that choices in Conc VI VI would be biased, relative to matching, in the direction that would optimize long-term reinforcement. Nonetheless, Houston and McNamara (1981) demonstrated that, with the exception of very extreme cases, overmatching optimizes rate of reinforcement in Conc VI VI, whereas most research reports undermatching patterns (Davison and McCarthy, 1988).

Another possibility is that matching is optimal at an

evolutionary scale, and thus the positive polarity of lean schedules improves the Darwinian fitness of the undermatching organism. Indeed, the notion that matching is innate has been advanced (Gallistel et al., 2007). Because this hypothesis is hard to falsify, we will not speculate in candidate mechanisms for the natural selection of undermatching.

A third possibility involves the misattribution of reinforcement. If reinforcers are occasionally attributed to the wrong schedule, it would be expected that most misattributions would favor the lean schedule, because most reinforcers are provided by the rich schedule (Davison and Jenkins, 1985). Reinforcement misattribution would thus yield undermatching. It is possible that the polarity of alternatives reflects the attribution of reinforcement: Positive alternatives would be those that are attributed more reinforcement than they actually yielded.

The misattribution hypothesis makes informative, verifiable predictions of choice behavior. For instance, if a few assumptions are made on how reinforcement operates on behavior that precedes the effective response (Catania, 1971; Killeen, 1994), misattribution explains why delays to changeover between alternatives decrease undermatching (Boelens and Kop, 1983; Shull and Pliskoff, 1967). It cannot, however, explain why overmatching happens at all (Baum et al., 1999; but see Wearden, 1983).

In synthesis, there is no clear rule yet for assigning polarity to alternatives. Assuming undermatching as the modal pattern of choice in Conc VI VI, the model of bipolar choice predicts an initial tendency to match choices to reinforcement, and the progressive acquisition of a tendency to allocate more behavior to the lean (positive) schedule. This latter tendency may be driven by long term contingencies, evolutionary predispositions, or by acquired errors in reinforcement attribution. Empirical research is necessary to establish whether bias in matching is acquired and why.

Prediction of Local Patterns of Choice

We have demonstrated that the global pattern described by the FS model may be derived from a simple choice mechanism that learns to "doubt" before selecting one of the alternatives but not the other. To the extent that a choice procedure yields the global pattern predicted by the FS model, be it a discrete choice, concurrent schedules, or concurrent chain schedules of any kind, the bipolar choice model can provide a plausible local choice mechanism. Can the same mechanism describe local patterns of choice? The data reported by Baum et al. (1999) provide an informative constraint to any model of local choice: Visit durations to the preferred (rich) alternative, but not to the non-preferred (lean) alternative, covary with relative rate of reinforcement. The algorithm described in Figure 2 would not yield this local pattern, mainly because it does not incorporate the burst-pause pattern of key-pecking that yields a high autocorrelation of choices in concurrent schedules (Nevin and Baum, 1980). This system may also need to incorporate MacDonall's (1999, 2000; MacDonall et al., 2006) insight that choice between concurrent schedules is constituted by choices between staying in and switching from each alternative. It is possible to conceive an undermatching organism that stays in the leaner alternative with a constant probability (X_S in Figure 2), but that occasionally reconsiders its choice of staying in the richer alternative, with a probability of actually leaving (ΔS) that negatively covaries with the rich rate of reinforcement. Such organism would behave like Baum et al.'s pigeons.

Conceivability does not entail necessity, but then again, no model is strictly necessary. Ours is no exception, but it may be argued that, without further empirical constraints, there are too many degrees of freedom in the model to be of any use. This does not ring true, because we have shown that local patterns of choice actually falsify a strict version of the model. A non-strict version of the model could be the one in which a choice is made with the probability X_S only when Equation 14 holds (Lefebvre, 2006); if this correlation is infringed, an organism begins restoring it by keeping the duration of visits to the lean alternative constant and increasing or decreasing the number of visits to the rich alternative as postulated in the bipolar model.

The model certainly needs to be developed to fully account for data but, more importantly, it also needs to make unique predictions that permit its empirical validation. Let us propose one prediction: If choice reconsideration is behaviorally expressed as orienting or moving toward the operandum without its actual activation, those actions should be evident in visits to one schedule—maybe at the end of response bursts—and not in visits to the other, and only during the acquisition phase. In Aparicio's (2001) barrier choice paradigm, we expect that, when short barriers separate the alternatives and undermatching is observed, rats would occasionally move toward the lever in the rich schedule before climbing away, whereas the abandonment of the lean schedule would be more resolute. If the height of the barriers is raised and overmatching occurs, we expect that incomplete motions would be evident in the lean, not in the rich schedule. Data on the topography of choice behavior is, unfortunately, scant. We hope that the model described here motivates further exploration.

Endnote

In Baum and colleagues' (1999) model, matching is discontinuous at $\log(B_R/B_L) = 0$ (preference indifference), and not necessarily when reinforcement rates are equal. This is because Baum and colleagues postulated their model as GML operating on the ratio of reinforcers obtained from the *preferred* and *nonpreferred* schedules. Equation 3 diverges from Baum and colleagues' to the extent that leaner schedules are preferred.

Acknowledgement

Federico Sanabria was supported by NIMH grant 1R01MH06686 to Peter R. Killeen. We thank William M. Baum, Peter R. Killeen, Nathalie Boutros, Douglas Elliffe, Alliston K. Reid, and two reviewers (one of them identified as Armando Machado) for their helpful comments on the draft of this paper, and Dr. Victorina Lefebvre without whose help this work would not be completed.

References

Adams-Webber, J. (1997). Self-reflexion in evaluating others. *Am. J. Psychol.* **110**, 527-541.

Aparicio, C.F., 2001. Overmatching in rats: The barrier choice paradigm. *J. Exper. Anal. Behav.* **75**, 93-106.

Aparicio, C.F., Baum, W.M., 2005. Fix and sample with rats and the dynamics of choice. *J. Exper. Anal. Behav.* **86**, 43-63.

Baum, W.M., 1974. On two types of deviation from the matching law: Bias and undermatching. *J. Exper. Anal. Behav.* **22**, 231-242.

Baum, W.M., 2002. From molecular to molar: A paradigm shift in behavior analysis. *J. Exper. Anal. Behav.* **78**, 95-116.

Baum, W.M., Rachlin, H.,1969. Choice as time allocation. *J. Exper. Anal. Behav.* **12**, 861-874.

Baum, W.M., Schwendiman, J.W., Bell, K.E., 1999. Choice,

contingency discrimination, and foraging theory. *J. Exper. Anal. Behav.* **71**, 355-373.

Boelens, H., Kop, P.F.M., 1983. Concurrent schedules - Spatial separation of response alternatives. *J. Exper. Anal. Behav.* **40**, 35-45.

Catania, A.C., 1971. Reinforcement schedules: The role of responses preceding the one that produces the reinforcer. *J. Exper. Anal. Behav,* **15**, 271-287.

Davison, M.C., Jenkins, P.E., 1985. Stimulus discriminability, contingency discriminability, and schedule performance. *Anim. Learn. Behav.* **13**, 77-84.

Davison, M. C., McCarthy, D., 1988. *The matching law: A research review*. New Jersey: Lawrence Erlbaum Associates.

De Villiers, P.A., Herrnstein, R.J., 1976. Toward a law of response strength. *Psychol. Bull.* **83**, 1131-1153.

Gallistel, C.R., King, A.P., Gottlieb, D., Balci, F., Papachristos, E.B., Szalecki, M., Carbone, K.S., 2007. Is matching innate? *J. Exper. Anal. Behav.* **87**, 161-199.

Herrnstein, R.J., 1961. Relative and absolute strength of response as a function of frequency of reinforcement. *J. Exper. Anal. Behav.* **4**, 267-272.

Herrnstein, R. J., 1982. Melioration as behavioral dynamism. In: M.L. Commons, R.J. Herrnstein and H. Rachlin (Eds). *Quantitative Analyses of Behavior. Matching and Maximizing Accounts* Vol. 2 (pp. 433–458). Cambridge, MA: Ballinger.

Heyman, G.M., Tanz, L., 1995. How to teach a pigeon to maximize overall reinforcement rate. *J. Exper. Anal. Behav.* **64**, 277-297.

Houston, A.I., McNamara, J., 1981. How to maximize reward rate on two variable-interval paradigms. *J. Exper. Anal. Behav.* **35**, 367-396.

Killeen, P.R., 1994. Mathematical principles of reinforcement: Based on the correlation of behavior with incentives in short-term memory. *Behav. Brain Sc.* **17**, 105-172.

Lau, B., Glimcher, P. W., 2005. Dynamic response-by-response models of matching behavior in rhesus monkeys. *J. Exper.*

Anal. Behav. **84**, 555-579.

Lefebvre, V. A., 1980. An algebraic model of ethical cognition. *J. Math. Psychol.* **22**, 83-120.

Lefebvre, V.A., 1982. *Algebra of conscience.* Dordrecht, Holland: D. Reidel.

Lefebvre, V.A., 1992. *A psychological theory of bipolarity and reflexivity. Lewiston,* NY: The Edwin Mellen Press.

Lefebvre, V.A., 2004. Bipolarity, choice, and entro-field. PROCEEDINGS. The 8th World Multi-Conference on Systemics, Cybernetics and Informatics. Vol. IV, 95-99.

Lefebvre, V.A., 2006. *Research on bipolarity and reflexivity.* Lewiston, NY: The Edwin Mellen Press.

Lobb, B., Davison, M.C., 1975. Performance in concurrent interval schedules: A systematic replication. *J. Exper. Anal. Behav.* **24**, 191-197.

MacDonall, J.S., 1999. A local model of concurrent performance. *J. Exper. Anal. Behav.* **71**, 57-74.

MacDonall, J.S., 2000. Synthesizing concurrent interval performances. *J. Exper. Anal. Behav.* **74**, 189-206.

MacDonall, J.S., Goodell, J., Juliano A., 2006. Momentary maximizing and optimal foraging theories of performance on concurrent VR schedules. *Behav. Process.* **72**, 283-299.

Nevin, J.A., Baum, W.M., 1980. Feedback functions for variable-interval reinforcement. *J. Exper. Anal. Behav.* **34**, 207-217.

Poulton E.C., Simmonds, D.C.V., Warren, R.M., 1968. Response bias in very first judgments of the reflectance of grays: Numerical versus linear estimates. *Percep. Psychophys.* **3**, 112-114.

Rachlin, H., 2000. *The science of self-control.* Cambridge, MA: Harvard University Press.

Sanabria, F., Baker, F., Rachlin, H., 2003. Learning by pigeons playing against tit-for-tat in an operant prisoner's dilemma. *Learn. Behav.* **31**, 318-331.

Shull, R.L., Pliskoff, S.S., 1967. Changeover delay and concurrent schedules - Some effects on relative performance measures. *J. Exper. Anal. Behav.* **10**, 517-527.

Staddon J.E.R, Motheral S, 1978. On matching and maximizing in operant choice experiments. *Psychol. Rev.*, **85**, 436−444.

Wearden, J.H., 1983. Undermatching and overmatching as deviations from the matching law. *J. Exper. Anal. Behav.* **40**, 333-340.